Scheduling Problems - New Applications and Trends

Edited by Rodrigo da Rosa Righi

Published in London, United Kingdom

IntechOpen

Supporting open minds since 2005

Scheduling Problems - New Applications and Trends
http://dx.doi.org/10.5772/intechopen.80171
Edited by Rodrigo da Rosa Righi

Contributors
Tahani Aladwani, Ade Jamal, Liliana Grigoriu, Hong Seong Park, Larysa Globa, Alexander Koval, Nataliia Gvozdetska, Volodymyr Prokopets, Surya Teja Marella, Thummuru Gunasekhar, Rodrigo Da Rosa Righi, Diorgenes Eugenio da Silveira, Cristiano Costa, Rodolfo Stoffel Antunes, Eduardo Souza dos Reis, Jorge Luis Barbosa, Marcio Miguel Gomes, Alvaro Machado Júnior, Rodrigo Simon Bavaresco, Rodrigo Saad

First published in London, United Kingdom, 2020 by IntechOpen
IntechOpen is the global imprint of INTECHOPEN LIMITED, registered in England and Wales, registration number: 11086078, 7th floor, 10 Lower Thames Street, London, EC3R 6AF, United Kingdom
Printed in Croatia

British Library Cataloguing-in-Publication Data
A catalogue record for this book is available from the British Library

Additional hard and PDF copies can be obtained from orders@intechopen.com

Scheduling Problems - New Applications and Trends
Edited by Rodrigo da Rosa Righi
p. cm.
Print ISBN 978-1-78985-053-6
Online ISBN 978-1-78985-054-3
eBook (PDF) ISBN 978-1-83962-169-7

For EU product safety concerns:
IN TECH d.o.o., Prolaz Marije Krucifikse Kozulić 3, 51000 Rijeka, Croatia, info@intechopen.com or visit our website at intechopen.com.

Meet the editor

Rodrigo da Rosa Righi is a senior member of both the Institute of Electrical and Electronics Engineers (IEEE) and Association for Computing Machinery (ACM). He is also a professor and researcher at the University of Vale do Rio dos Sinos (Unisinos), Brazil. Dr. Righi concluded his post-doctoral studies at the Korean Advanced Institute of Science and Technology (KAIS), South Korea, in the subjects of the Internet of Things (IoT) and cloud computing. He is a coordinator of national and international projects in the areas of resource management in distributed systems, fog and cloud computing, Industry 4.0, and artificial intelligence. His research interests include performance analysis, predictive maintenance, event prediction and correlation, and cloud and fog resource elasticity. More details about his research can be found at: http://professor.unisinos.br/rrrighi.

Contents

Preface

Scheduling is defined as the process of assigning operations to resources over time to optimize a criterion. The requirements for scheduling mentioned in the literature include the minimization of several factors, including completion time of the set of services under consideration (the makespan), mean Work in Process (WIP), mean manufacturing time (the mean flow time), mean delay, and mean processing cost, and the maximization of productivity.

Scheduling is essential in many different fields, including e-health, high-performance computing, data science, big data, and Industry 4.0. Our book covers new aspects and uses of scheduling and load balancing, detailing challenges and new trends in the field. In three sections ergo seven chapters, we revisit the concepts of scheduling and the novelties of scheduling problems, in addition to examining new areas that are benefiting from these concepts to both improve efficiency and reduce costs.

Scheduling has a broad impact on several areas. Considering this, the content of this book is not limited to engineering, but also covers other areas such as biological, chemical, and computational fields. Thus, this book will be of interest to those working in the decision-making branches of production in various operational research areas as well as in the design of computational methods. People from diverse backgrounds like academia, industry, and research can take advantage of this volume.

Prof. Dr. Rodrigo da Rosa Righi
Professor and Researcher in the Applied Computing Graduate Program (PIPCA),
Universidade do Vale do Rio dos Sinos (UNISINOS),
São Leopoldo, Brazil

Section 1

New Scheduling Approaches and Algorithms

Chapter 1

Global Optimization Using Local Search Approach for Course Scheduling Problem

Ade Jamal

Abstract

Course scheduling problem is a combinatorial optimization problem which is defined over a finite discrete problem whose candidate solution structure is expressed as a finite sequence of course events scheduled in available time and space resources. This problem is considered as non-deterministic polynomial complete problem which is hard to solve. Many solution methods have been studied in the past for solving the course scheduling problem, namely from the most traditional approach such as graph coloring technique; the local search family such as hill-climbing search, taboo search, and simulated annealing technique; and various population-based metaheuristic methods such as evolutionary algorithm, genetic algorithm, and swarm optimization. This article will discuss these various probabilistic optimization methods in order to gain the global optimal solution. Furthermore, inclusion of a local search in the population-based algorithm to improve the global solution will be explained rigorously.

Keywords: course scheduling, optimization, local search, genetic algorithm, particle swarm optimization, combinatorial optimization problem, probabilistic optimization algorithm

1. Introduction

Scheduling is the process of assigning a set of given tasks to resources by some means. Among the resources, time resource usually plays a central role in scheduling process; hence this process is often called timetabling. Besides the time resource, there are other resources involved in the scheduling process such as space or room, machine or tools, and human resources. The resources are usually subject to constraints that make scheduling problems interesting for researchers in finding an optimal solution or in developing a method for solving it. Course scheduling problem attracts researchers from the field of operation research and artificial intelligence [1–9].

This manuscript will focus on the problem of university course scheduling which has several variants such as school timetabling [10] and examination scheduling [11–13]. The variation of course scheduling problem is merely due to different constraints on the resources involved in the scheduling processes. Despite of these variations, they can be considered as the same family of course scheduling problem. In the scheduling problem, courses or exams have to be assigned into time and space resources by considering some constraints. University course scheduling problem can be divided into two categories, post-enrolment scheduling [1–4] and prior-enrolment scheduling [5–9]. In the prior-enrolment-based course scheduling, students are not taken into account as an individual person but as a group of study curriculum and student grade; hence it is also named a curriculum-based scheduling [5, 6]. In the post-enrolment-based course scheduling, students and faculty members or lecturers are considered as individual person and not as specified parameter on courses.

University course scheduling problem is simple to understand, yet complex enough to admit solution at varying level of difficulty in the implementation. Several studies of university course scheduling are conducted using operation research, human computer/machine interface, and artificial intelligence. The main issues in the solving method of university course scheduling problem are quality of the schedule solution, namely, the optimal solution, and time spent to produce the schedule solution, i.e., algorithm efficiency.

The most traditional technique for solving the course scheduling problem is the graph coloring technique [14, 15]. Graph coloring algorithm comes from a classical problem in graph theory which implies the problem to color the nodes of an undirected graph such that no two adjacent nodes share the same color. The course scheduling problem is modeled by letting the nodes and edges represent the courses and the common students, respectively. Dandashi and Al-Mouhamed [15] have given a good historical review on the graph coloring technique since 1967 up to recently.

The second group of solution method is the family of local search methods. This is a heuristic-based search method. It finds the best solution among a number of candidate solutions by applying local change from the last found solution. This is a very fast algorithm, but it has a downside that it can easily be stuck at a local optimum. This local optimum issue can be cured by many local search variants such as the taboo search [13], simulated annealing search [2, 10, 12], and improved hill-climbing search [1, 4, 9, 11, 16].

The third group of solution method is the population-based optimization methods that gain more attention by researchers in seeking new methods that are inspired by the nature phenomena such as genetic algorithm [7, 8, 12], evolutionary algorithm [3, 17, 18], ant colony algorithm [19], bee colony algorithm [20, 21], firefly algorithm [22], and particle swam optimization [23, 24].

Evolutionary algorithm was originally not a population-based approach as introduced by Rechenberg in 1965 where only one species is mutated and only one species, i.e., the fittest one, survived in every evolution generation. Mutation is the only reproduction mechanism necessary in the evolutionary algorithm. Crossover mechanism is another reproduction mechanism inspired by biological evolution theory. Crossover mechanism is a simplification model of genetically offspring from mating process of a parent pair. The work of John Holland in the early 1970s included both genetic operators, and since then a so-called genetic algorithm became popular that belongs to the evolutionary strategy family. In genetic algorithm, each individual in a population forms a candidate solution. The candidate solution is evolved by mutation and crossover mechanism in every generation. Through fitness selection scheme, they move toward a better generation.

After successful mimicking of the nature phenomena from the evolution theory, once again, Mother Nature has inspired researchers to develop new optimization algorithm based on swarm intelligent theory. Swarm behavior or swarming is a collective behavior exhibited by particularly animals which aggregate together in finding food and moving or migrating in some direction. Ant colony optimization algorithm is one of the first metaheuristic optimizations in this group of optimization method, initially proposed by Marco Dorigo in 1992. Initially ants wander randomly, and upon finding food they return to their colony while leaving trail called pheromone trails. When other ants find such trail, they are likely to follow the trail and return and reinforce the trails if they eventually find food. The pheromone trail is the main issue of swarm intelligent communication in ant colony, on which the algorithm was developed.

Another algorithm that imitates the intelligent foraging behavior of animal is artificial bee colony optimization algorithm proposed by Karaboga in 2005. There are three groups of bees in bee colony, i.e., employed bee, scouts, and onlookers. Employed bees go to their food source and back to hive and dance. Onlookers watch the dances of the employed bees, and depending on employed bee's waggle dances, food sources are chosen or abandoned. The employed bees whose food sources have been abandoned become a scout and start to seek for a new food source.

The most recent bio-inspired algorithm, as far as the author's knowledge, is the firefly algorithm developed by Xin-She Yang in 2008. It is a heuristic algorithm which is a population-based stochastic method which is derived and motivated by the flashing or mating behavior of fireflies. The position of all fireflies represents a possible set of solutions, and their light intensities represent corresponding fitness values or quality of all solutions.

Particle swarm optimization is a population-based evolutionary computation technique developed by Eberhart and Kennedy in 1995, inspired by social behavior of bird flocking or fish schooling. This algorithm is the simplest model of swarm behavior algorithm. This algorithm shares similarity with genetic algorithm, but it differs mainly due to the absence of genetic operator. A kind of communication between particles in the swarm controls the movement of each particle in searching food. When an animal spots a location that is rich of food, it memorizes the location until better location is found. The movement of each particle is calibrated to its best location so far and the best location from the whole animals in the swarm. The algorithm is also much simpler because it has only few parameters to adjust compared to genetic algorithm. Its simplicity and its generic computation model for broad application make this algorithm more attractive to assess than other new algorithms inspired by animal behavior as reviewed by Poli et al. [23].

The following sections will be structured as follows: first we will discuss the probabilistic optimization methods from three different previously described approaches. Thereafter we will rigorously explain the university course scheduling problem and how we model it appropriately for all the employed solution methods. Discussion of experimental result will be presented in the last section before some concluding remarks are briefly inscribed.

2. Probabilistic optimization method

From the beforehand described optimization approaches for discrete problem such as university course scheduling problem, we can summarize that there are three groups of solution approach, namely, coloring graph, local search approach, and the population-based approach. The population-based approach can be further classified into the population-based evolutionary approach and the population-based social behavior approach.

While the probabilistic characteristic is inherent in the population-based approaches, the local search approach is a single-based solution technique and normally not a probabilistic solution method. In this article, we will bring probabilistic nature into the local search approach by introducing scattered neighborhood [9] and multiple random start local search method [25]. We will discuss thoroughly this multiple-scattered local search and compare these two population-based methods, namely, genetic algorithm and particle swarm optimization, respectively, from evolutionary approach family and social behavior approach family.

2.1 Multiple-scattered local search (MSLS)

Local search approach has little probabilistic nature. In its original form, namely, the steepest-ascent hill-climbing method, wherein from its current position systematically for every possible single mutation is evaluated to find the highest fitness increase in the neighborhood, this method has no probabilistic aspect at all. The fast local search, called the next ascent hill-climbing [26], wherein a single mutation is evaluated systematically from current position until any increase in the fitness is found, has a little probabilistic nature. Random mutation hill-climbing [26] or scattered local search [9] is a probabilistic variation of local search wherein from its current position, a random mutation mechanism takes place until an increase in fitness is found. Random mutation hill-climbing takes randomly a single mutation, while scattered hill-climbing takes a certain number of mutation randomly, and the highest increase in fitness is chosen. Other local searches, such as taboo search and simulated annealing search, are also improvements and derivatives of the hill-climbing search method. However, they can hardly be classified as hill-climbing search technique since the changing makes them differ too much from the original method.

In general, the hill-climbing procedures are as follows:

1. Evaluate the fitness, i.e., the heuristic objective function, for the current position.

2. Find the candidate for the next position in the neighborhood area by small changing from the current position and then evaluate the fitness for the candidate position.

3. If there was an increase in fitness, then set the candidate for the next position as the current position.

4. Repeat from step 1 until the maximum iteration was reached or other stop criteria are fulfilled.

The difference between various hill-climbing search described in previous paragraph is found only in step 2 of the above procedure. Because of systematically searching for the candidate for the next position, the steepest ascent hill-climbing and the next ascent hill-climbing methods are classified as deterministic search techniques. The random mutation hill-climbing and the scattered hill-climbing are probabilistic search techniques due to random mutations in searching for the next position.

Random restart hill-climbing is another approach to embed the probabilistic nature. In case the hill-climbing search stopped due to its maximum iteration, the whole hill-climbing search above restarted with a completely new initial position which set randomly. The restart algorithm will continue until the optimization criteria are reached. The multiple-scattered local search is actually a set number of scattered hill-climbing searches running with randomly set initial positions.

Let np be the number of initial position and n the number of scattered candidate position in the neighborhood of current position, and then the algorithm of the multiple-scattered local search is as follows:

1. Set up *np* initial positions randomly, and evaluate their heuristic fitness or objective functions, and save in array of *np* current positions.

2. Repeat the following steps until maximum iteration or another stop criterion is reached:

    a. For each position in a set of *np* positions, do:

        i. Find *n* new positions in the neighbor of current position.

        ii. Evaluate the fitness of each new position, and save the best position whose fitness has the highest value.

        iii. If the best position gives an increase in fitness compared to the current position, set the best position as current position.

2.2 Genetic algorithm

Genetic algorithm is a population-based optimization technique that simplifies survival of the fittest principle from the evolution theory. Any individual chromosome represents a solution state of the problem. Every generation has a population that consists of a fixed number of individual instances. The population of the next generation will inherit their ancestor by crossover of two mating chromosomes, by mutation of single individual chromosome, and by elitism. All of these three evolution mechanisms hold the true survival of the fittest principle.

Elitism in the evolution mechanism means superior chromosomes survive to the next generation without any changes. This elite group which is only a small fraction of the population has the highest fitness in the population. The rest of the population will be selected in pairs through a probabilistic scheme involving their fitness values. Each pair acts as parent that will have two descendants produced by crossover between two parent's chromosomes. Since the population size is constant, every pair of children replaces their parents in every generation. Finally, mutations take place in all children by chance of probability distribution.

The following is the pseudocode of genetic algorithm:

1. Set up *np* individual chromosomes for the first generation.

2. Repeat the following steps limited by maximum generation number:

    a. Evaluate the fitness values of every chromosome in the population of current generation.

    b. Sort the chromosomes by the fitness values.

    c. Select a small number of *ne* elite chromosomes, i.e., the highest ranked chromosomes.

    d. Select a pair of parents via a selection mechanism from the rest of the chromosomes.

    e. Based on a crossover probability, crossover scheme takes place between each selected pair, and the two successors will replace the parents.

f. Based on a mutation probability, every single chromosome outside the elite group undergoes a mutation.

Elitism mechanism has a purpose to make sure that the individuals with the highest fitness do not vanish by chance of probabilistic. Pairing the parents aims to find better descendants which differs from both parents but still inherits their characteristics. Hence, crossover mechanism has a function of an exploration for new chromosomes. Single chromosome is slightly changing by mutation which aims to improve the individual in exploitation scheme. Exploration power of crossover combined with exploitation power of mutation and an additional power, a conservatism of elitism, makes genetic algorithm very popular for a long time in the area of artificial intelligence and organization research for optimization problem [17].

2.3 Particle swarm optimization

Foraging behavior of some animals is in a group of numerous individuals in swarm formation. In the swarm behavior, animals such as birds, insects, or fishes move in such mechanism that they do not collide with each other, but they can cover broaden area for finding the foods. This behavior inspired an optimization technique called particle swarm optimization which is rather simple than the previously nature-inspired genetic algorithm. Every position of animals in the swarm represents a solution state of the problem. Every animal or particle moves in directions that are influenced by the best individual position so far and the best position of all individuals in the swarm. The best position means, in the real natural condition, the richest food available, and in the optimization problem means the highest fitness solution state.

Let *np* particles be in the swarm and each particle has its own initial positions ***Xo*** and velocity ***Vo***; then, the pseudocode of the particle swam optimization is as follows:

1. Evaluate the fitness value of each initial position ***Xo***, and save the individual best position as ***XPbest*** and the global best position as ***XGbest***.

2. Repeat the following steps until maximum iteration or another stop criterion is reached.

   a. Update the velocity of each particle for the next move as follows:

   $$Vi + 1 = Vi + r_1 * (XPbest - Xi) + r_2 * (XGBest - Xi)$$

   b. Update the new position of each particle as follows:

   $$Xi + 1 = Xi + Vi + 1$$

   c. Evaluate the fitness value of the new position of each particle, and update the individual best position and the global best position if necessary.

According to James Kennedy who proposes particle swarm optimization (1995), the adjustment particle movement toward the individual and global best solution is conceptually similar to the crossover mechanism in genetic algorithm. This movement ensures exploration power of the algorithm. The two types of best positions are in some sense acting as elitism in genetic algorithm which holds the conservatism of the particle swarm optimization algorithm.

2.4 Exploitation power of local search in the global optimization

Genetic algorithm and particle swarm optimization have great power of exploration. The exploration power could almost guarantee finding a global optimum if the number of iteration is large enough or at least finding a sufficient global optimum for a reasonable number of iteration. Lack of exploitation power in particle swarm optimization and slight exploitation power in genetic algorithm make these two algorithms slow in finding of good enough solutions. We will introduce the exploitation power into genetic algorithm and particle swarm optimization with the aid of the local search on the elite groups [17, 18].

2.4.1 Hybrid genetic algorithm (HGA)

In the original genetic algorithm, the elites are not mutated, but could be replaced by new elites in the next generation. Empowering the algorithm using single iteration of scattered local search on the elites, the hybrid genetic algorithm becomes as follows:

1. Set up *np* individual chromosomes for the first generation.

2. Repeat the following steps limited by maximum generation number:

   a. Evaluate the fitness values of every chromosome in the population of current generation.

   b. Sort the chromosomes by the fitness values.

   c. Select a small number of *ne* elite chromosomes, i.e., the highest ranked chromosomes.

   d. For each chromosome in *ne* elite, do:

      i. Randomly develop *n* new mutated chromosomes.

      ii. Evaluate the fitness of each new mutated chromosome, and save the best chromosome whose fitness has the highest value.

      iii. If the best chromosome gives an increase in fitness compared to the current elite chromosome, set the best chromosome as the new elite chromosome.

   e. Select pair of parents via a selection mechanism from the rest of the chromosomes.

   f. Based on a crossover probability, crossover scheme takes place between each selected pair, and the two successors will replace the parents.

   g. Based on a mutation probability, every single chromosome outside the elite group undergoes a mutation.

2.4.2 Hybrid particle swarm optimization (HPSO)

We will bring the exploitation power in the particle swarm optimization by performing scattered local search on the individual best positions and the global best position of particles in the swarm. Hence, the hybrid particle swarm optimization algorithm becomes as follows:

1. Evaluate the fitness value of each initial position ***Xo***, and save the individual best position as ***XPbest*** and the global best position as ***XGbest***.

2. Repeat the following steps until maximum iteration or another stop criterion is reached.

   a. Update the velocity of each particle for the next move as follows:

      $$Vi + 1 = Vi + r_1 * (XPbest - Xi) + r_2 * (XGBest - Xi)$$

   b. Update the new position of each particle as follows:

      $$Xi + 1 = Xi + Vi + 1$$

   c. Evaluate the fitness value of the new position of each particle, and update the individual best position and the global best position if necessary.

   d. For each individual best position ***XPbest*** and the global best position ***XGbest***, do:

      i. Find *n* new positions in the neighbor of the best position.

      ii. Evaluate the fitness of each new position, and save the new best position whose fitness has the highest value.

      iii. The new best position will replace the best position under consideration, if it gives increase in fitness.

3. University course scheduling problem and model

3.1 Problem definition

University course scheduling is a process of assigning a set of courses to limited time resources and space resources, namely, classrooms. We consider only curriculum-based or prior-enrollment course scheduling. In this case, each course has been set who will teach and which student group will attend the course. The assigning process must satisfy some hard constraint, for instance, avoiding lecturer conflict. In addition, some soft constraint such as considering preference time of lecturer will be desired to be fulfilled in the schedule.

Formally, the curriculum-based course scheduling problem will be described as follows:

- There is a set of courses [C_1, C_2, ... C_{nc}] where each course is attended by a number of students from specified studies and taught by specified lecturers. A course could be given in more than 1 course hour and probably need computer equipment in the classroom.

- There is a set of lecturers [L_1, L_2, ... L_{nl}] who have been assigned to teach some courses and probably have some unavailable time slots and some preference time slots to teach.

- There is a set of student group [S_1, S_2, ... S_{ns}] which consist of a number of student from the same department and the same grade.

- There is a set of room [R_1, R_2, ... R_{nr}] which has a number of seat capacity and some room equipped with computers.

- There is a set of time slot [T_1, T_2, ... T_{nt}] which can be expressed in daytime and clock time.

A feasible course schedule has to satisfy the following eight hard constraints:

- Complete schedule which means all courses have been assigned into available time and rooms

- Avoiding room conflict which means no room is used by more than one course event at the same time slot

- Avoiding lecturer conflict which means no lecturer teaches more than one course event at the same time slot

- Avoiding student conflict which means no student group from the same department and same grade must attend more than one course event at the same time

- Avoiding excessive attending student number in classroom

- Avoiding lecturers' unavailable time which means no lecturer teaches in his or her unavailable time slot

- Proper equipped classroom which means room has a feature needed by the assigned course

- Continuous course event which means multiple hour courses must be assigned in the same room contiguously

Beside the hard constraints that must be fulfilled unconditionally, course schedule preferably satisfy some soft constraints such as:

- Avoiding lecturer's preventive time which means lecturers do not teach in his or her preventive time slot.

- Lecturer's preference time which means lecturers teach in their preference time

- Full classroom which means the number of students joining the course is more than the half of classroom capacity

A course schedule is defined as optimum if the number of hard constraint violation is zero and the number of soft constraint fulfillment is as many as possible.

3.2 Course scheduling model

Computational model of a course schedule can be represented as a two-dimensional matrix where rooms and time slot denote as row number and column number [3, 5, 7]. Each matrix cell is filled up with course event. Time slots comprise of day and hour, for instance, if the number of day is 5 and every day consists of 8 course hours, then the number of time slot columns are 40 time slots. A variant of this model has flexible time slot length [7].

A slightly different model from the two-dimensional model is a three dimensional-model wherein the two components of time slot are put into a two-dimensional matrix, namely, hour in row and day in column. The room dimension is placed in the third direction of matrix [9, 17, 18, 27]. As in the two-dimensional matrix model, every matrix cell contains a single hour of course event.

These two matrix-based models have an advantage that the "room conflict" hard constraint is inherently fulfilled. However, it has disadvantage on "complete schedule" hard constraint, namely, ensuring all course events are completely put in the model. Hence the matrix-based model is inappropriate for randomly assigning course events in the matrix. Another problem with the matrix-based model is in crossover evolution mechanism where parent mating could make a course event duplicate and course event missing in their successors [9, 17].

To get around these two problems, we will use a list of 3-tuple which consists of course event, room, and time slot $< C_i, R_j, T_k >$ [6, 25]. This 3-tuple list can be simplified by introducing a space–time function f(R_j, T_k). If we denote all the variable using only their indices in the 3-tuple such that $< C_i, R_j, T_k >$ is written as $< i, j, k >$, then the space–time function can be expressed as

$$f(j,k) = k + (j - 1) * nt \tag{1}$$

where $j = 1 .. nr$ is room index and $k = 1 .. nt$ is time slot index and maximum value of $f(j,k) = nr^{*}nt$.

Hence the 3-tuple list schedule is shortened as a vector *Sch*:

$$Sch = \{f_1(j,k), f_2(j,k), \ldots, f_i(j,k), \ldots f_{nc}(j,k)\} \tag{2}$$

where index i represents a course which is associated to student groups S_l, lecturers L_m, and course event duration.

The first hard constraint, namely, the complete schedule constraint, is assured by schedule model in Eq. (2). The other seven hard constraints are taken into account in penalty function *HC(Sch)* which is an accumulation of every hard constraint violation in each course C_i. The three soft constraints are taken into account in score function *SC(Sch)* which is an accumulation of every soft constraint fulfillment in every course C_i. Hence the fitness function is formulated as

$$fitness(Sch) = w_{sc} * SC(Sch) - w_{hc} * HC(Sch) \tag{3}$$

where w_{sc} and w_{hc} are weight factors. The objective of optimization is to maximize this fitness function and make sure a zero hard constraint penalty function *HC(Sch)*.

Using the tuple list scheduling model, the problem size of course scheduling is mainly determined by the number of course *nc*.

3.3 Computational model

In the previous section, we explain how the scheduling model is described including the heuristic fitness function. Here we will describe the computational model of three algorithms, namely, multiple-scattered local search, hybrid genetic algorithm, and hybrid particle swarm optimization. Computational model identifies parameters for the computational size. The objective of defining the computational model is to ensure a fair comparison among three different types of algorithm, because considering only an equal problem size can yield biased evaluation if the computational model is significantly different.

The global search computational model is governed by population number *np*. In the case of multiple-scattered local search, this exploration power size determines the number of hill climbers; in the case of hybrid genetic algorithm, it is the amount of chromosomes; and for hybrid particle swarm optimization, it is set by the number of particles in the swarm. The local search algorithm of all three algorithms implements the same scattered local search; hence the size of this exploitation power is determined by the number of probable positions in the neighborhood *ne*.

The computational model size, i.e., population number *np* and neighbor number *ne*, and the problem size, i.e. course number *nc*, will completely govern the size of three different algorithms. The performance of the success runs for the heuristic approach control by the required number of iterations or generation in the case of genetic algorithm.

4. Experimental results and discussion

4.1 Evaluation model

Because course scheduling is a non-deterministic polynomial complete problem (NP-problem) [5, 28], algorithm evaluation tool for a deterministic polynomial problem (P-problem), such as complexity asymptotic analysis, is not really appropriate to evaluate course scheduling algorithms. An empirical approach was introduced by Hoos [29, 30] to analyze the behavior of non-deterministic polynomial problem for such three algorithms under consideration. Run time distribution (RTD) and run length distribution (RLD) are empirically constructed by running the same algorithm with the same condition for a sufficient number of runs until some stop criteria are reached or up to some cutoff time or maximum iteration. For each run, the required run time and the required number of iteration, for RTD or RLD, respectively, to reach a good solution are recorded.

RTD and RLD will represent cumulative probability distribution function:

$$F(x) = P\,|\,x \leq X\,| \tag{4}$$

where X is a random variable which represents required run time in RTD or required run length in RLD.

Specification	Set I	Set II
Number of courses	25	51
Number of rooms	2	4
Number of time slots	80	160
Number of instructors	14	23
Number of student group	4	7
Number of course hours	67	138

Table 1.
Two sets of small course scheduling data [15].

4.2 Experimental data

We will evaluate the three algorithms using two sets of small curriculum-based course scheduling problems as given in **Table 1**.

On each set, evaluation will be performed for 250 runs on each set of scheduling problems until at least one zero *HC(Sch)* or a specified maximum iteration is reached. Size of computational model of all three algorithms, i.e., population number *np* and neighborhood number *ne*, will be varied to grant the performance behavior of these three different algorithms.

4.3 Discussion of results

The effect of two computational size parameters, namely, the population number *np* and the neighborhood number *ne*, will be studied. While the population number *np* gives the exploration force to enhance a global search, the neighborhood number *ne* provides the exploitation force for the local search.

4.3.1 Exploration for global search

We will firstly investigate influences of parameter *np* on the success probability for each algorithm separately using the scheduling problem set I. Test results from the hybrid particle swarm optimization are represented in **Figures 1** and **2**, respectively, RLD and RTD. **Figure 1** shows that the higher number of population, the better its run length distribution, namely, the most left distribution function. It is also shown that using cutoff of 1000 iterations, 100% probability of success to find a zero *HC(Sch)* is obtained for a population number *np* of at least 40, and for the smaller *np*, it is slightly less than 100%. However, the variation of population number almost does not affect RTD for the hybrid particle swarm optimization as shown in **Figure 2**.

Figures 3 and **4**, respectively, represent RLD and RTD from the multiple-scattered local search. **Figure 3** shows the same behavior as **Figure 1** for the hybrid particle swarm optimization, namely, increasing number of population shifts RLD function to the left side, the better behavior. Run time distribution using multiple-scattered local search depends significantly on the population number because the

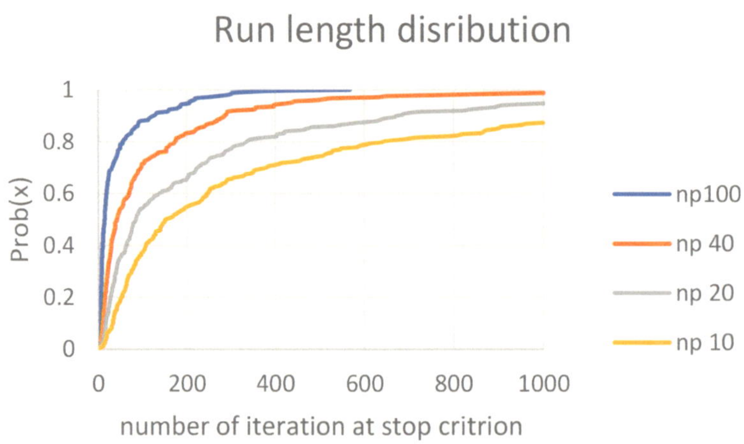

Figure 1.
RLD resulting from hybrid particle swarm optimization for population number np 10, 20, 40, and 100 and neighborhood number ne 20.

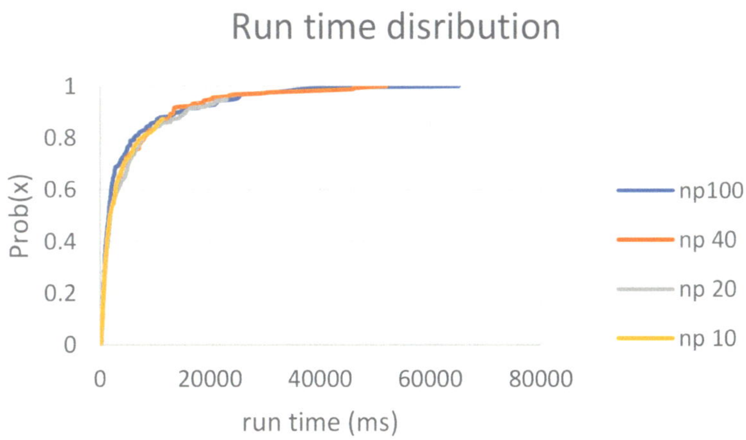

Figure 2.
RTD resulting from hybrid particle swarm optimization for population number np 10, 20, 40, and 100 and neighborhood number ne 20.

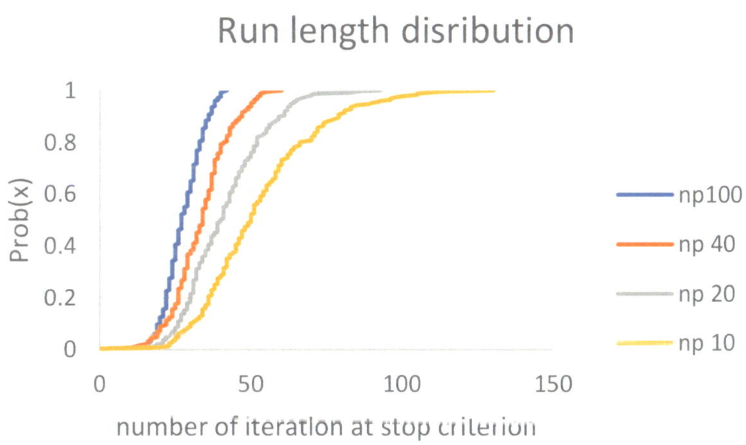

Figure 3.
RLD resulting from multiple-scattered local search for population number np 10, 20, 40, and 100 and neighborhood number ne 20.

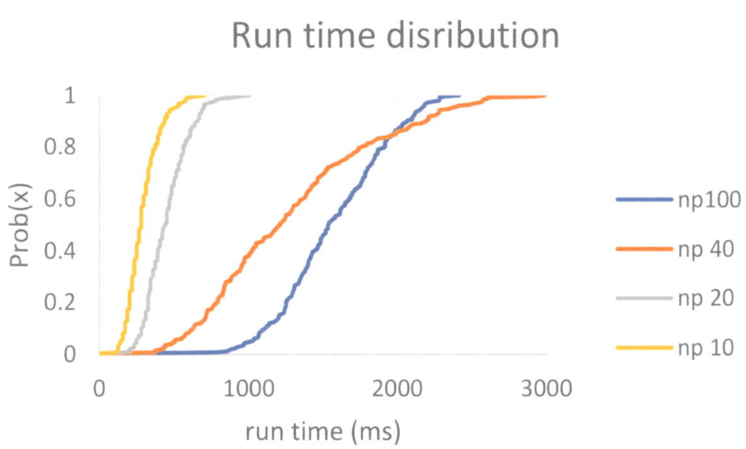

Figure 4.
RTD resulting from multiple-scattered local search for population number np 10, 20, 40, and 100 and neighborhood number ne 20.

required run time growth is higher than the reduction of required run length as function of population number as shown in **Figure 4**.

Reducing the number of population until *np* = 10, we found that using multiple-scattered local search, finding of a feasible schedule is assured if one lets the algorithm run up to 1000 iterations. We have investigated thoroughly this algorithm for lower population number until it becomes a single-scattered local search as presented in **Figures 5** and **6**. Note that we set maximum iteration of 3000 for this test. We found that a success run probability of 100% at maximum iteration of 3000 needs at least *np* = 4 population members and the minimum population number is getting higher, i.e., *np* = 8, for larger schedule problem, i.e., scheduling problem set II as shown in **Figure 6**.

Figures 7 and **8** show the result of hybrid genetic algorithm. In this case, increasing number of population *np* merely improves the run length distribution as shown in **Figure 7**. **Figure 8** shows that the increasing number only made things worse, as logically higher population number needs more run time for the same level of probability.

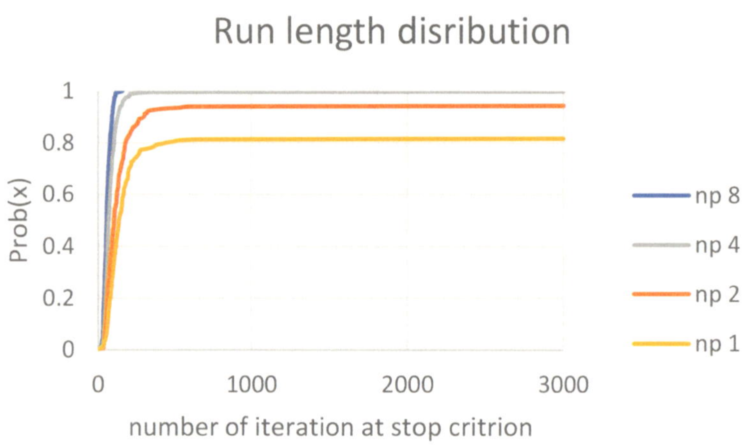

Figure 5.
RLD resulting from multiple-scattered local search for population number np 1, 2, 4, and 8 and scheduling in two rooms.

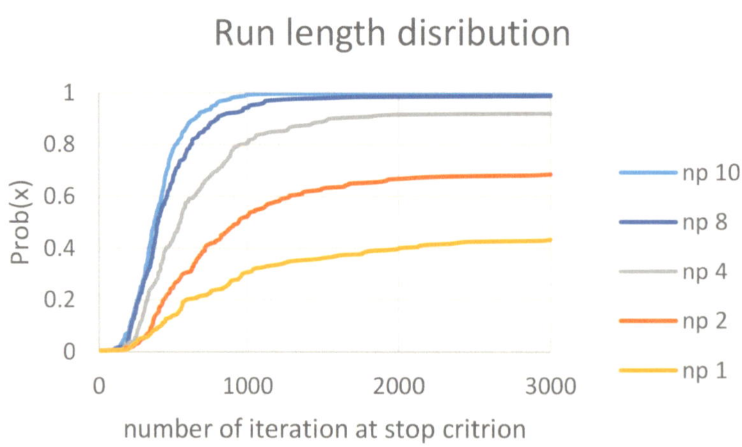

Figure 6.
RLD resulting from multiple-scattered local search for population number np 1, 2, 4, 8, and 10 and scheduling in four rooms.

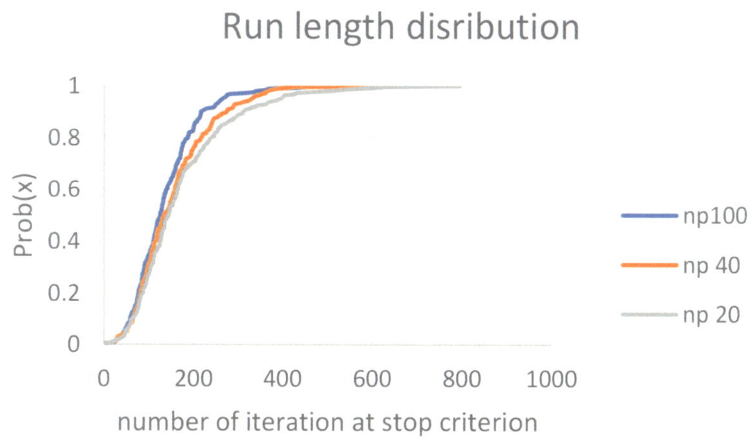

Figure 7.
RLD resulting from hybrid genetic algorithm with population number np 20, 40, 100 and neighborhood number ne 20.

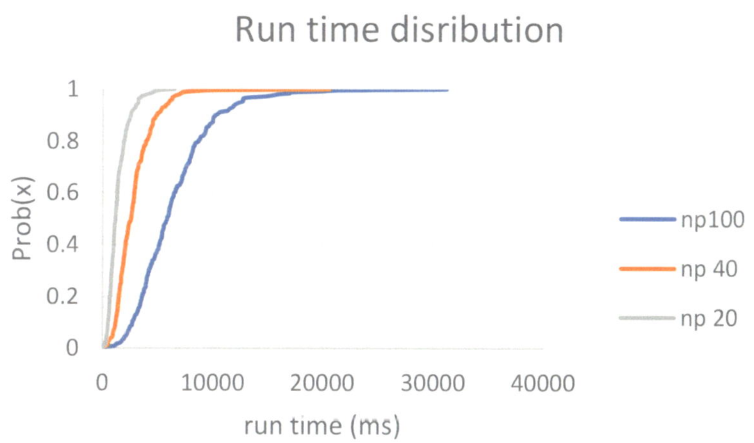

Figure 8.
RTD resulting from hybrid genetic algorithm for population number np 20, 40, and 100 and neighborhood number ne 20.

4.3.2 Exploitation for local search

The novelty of the presented hybrid algorithms is the introducing of exploitation force in the original algorithm where exploration power is essential. This enhancement is affected by the number of scattered candidates in the neighborhood of the elites, denoted as *ne*. If this neighborhood number *ne* is zero, it means the hybrid algorithm is exactly the same as the original versions.

Figure 9 depicts effect of local search in the hybrid genetic algorithm on the run length probability distribution. Even a small neighborhood number $ne = 2$ affects the probability function significantly compared to the original genetic algorithm.

Figure 10 depicts effect of local search in the hybrid particle swarm optimization on the run length probability distribution. Using 6000 iterations as the maximum stop criterion, the original particle swarm optimization fails to yield any feasible solution; hence no result is given in **Figure 10** for the original version. Even a small neighborhood number $ne = 2$ affects the probability function significantly compared to the original genetic algorithm.

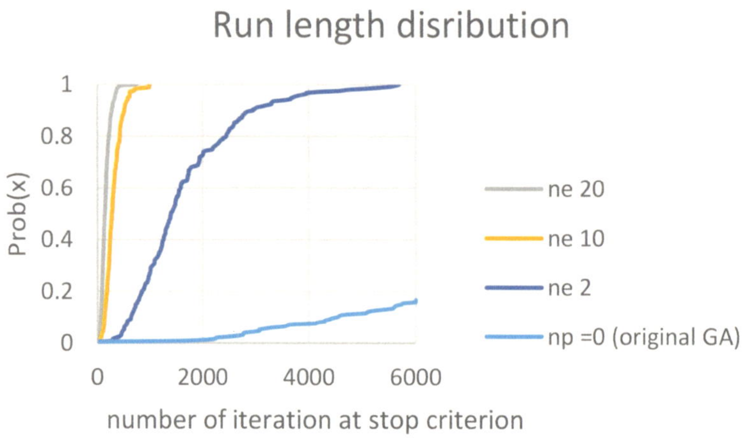

Figure 9.
RLD resulting from hybrid genetic algorithm with variation of neighborhood number ne 2, 10, and 20 and the original genetic algorithm.

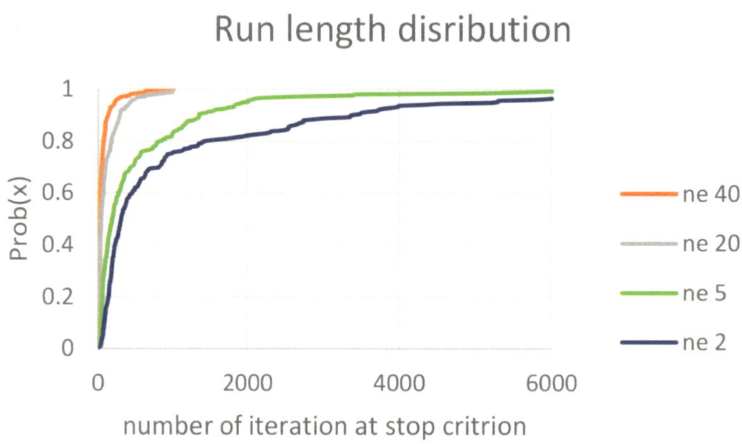

Figure 10.
RLD resulting from hybrid particle swarm optimization with variation of neighborhood number ne 2, 5, 20, and 40.

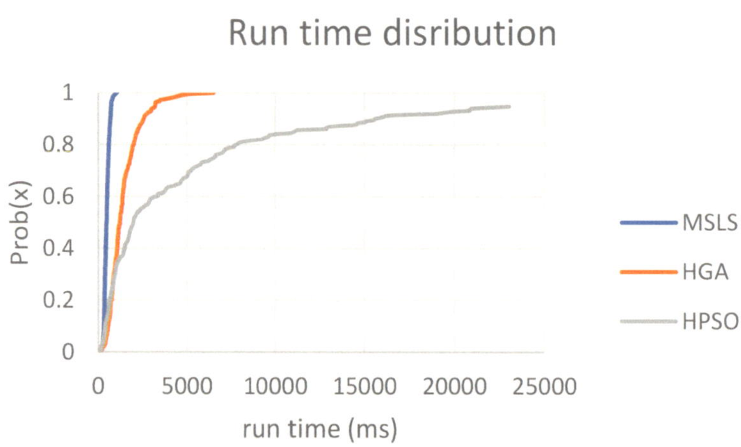

Figure 11.
RTD resulting from the three algorithms for population number np 20 and neighborhood number ne 20, which are multiple-scattered local search, hybrid genetic algorithm, and hybrid particle swarm optimization.

4.3.3 Comparison between three algorithms

Figure 11 depicts RTD for three algorithms for the same computational size, i.e., $np = 20$ and $ne = 20$. It shows that multiple-scattered local search is the most efficient in time and also in the required iteration number, while hybrid particle swarm optimization is the worst.

5. Conclusions

Course scheduling problem is a discrete optimization problem and considered as NP complete problem which is hard to solve. Therefore, we presented three algorithms which are modification from three popular algorithms. The first algorithm is multiple-scattered local search which is an enhancement of the hill-climbing search. The improvement is achieved by introducing exploration search power in the original single local search.

The hybrid genetic algorithm and the hybrid particle swarm optimization are the last two presented algorithms. The original version of these two algorithms is well known for their powerful of exploration ability in searching of global solution for a sufficiently large number of iterations. The hybrid enhancement of these two algorithms was implemented by implanting scattered local search on the small elite group. The enhancement aimed to accelerate searching process.

Course scheduling problems used for the evaluation of three algorithms are curriculum-based or pre-enrollment course schedules which must fulfill eight hard constraints and three soft constraints. The course schedule problem was modeled using a list of 3-tuple which consists of course event, room, and time slot $<C_i, R_j, T_k>$ which further simplified as a vector containing time–space allocation of every course. The essential advantage of using this course model is to resolve the problem of missing courses and double allocated of the same courses in the evolution mechanism. Furthermore, this course model can be used for all three algorithms under consideration with comparable computation model size, i.e. the population number *np* the neighborhood number *ne*.

The experimental test results have proven that the population number consistently governs the exploration power for better global search in term of required iteration numbers at cost of more time needed. In the case of multiple-scattered local search, only a small number of population are needed to obtain a good probability behavior of success run.

The effect of implanting local search in two originally global search algorithms is more remarkable. Letting a very small-scattered local search in the elite group has improved the cumulative probability distribution function of hybrid genetic algorithm and hybrid particle swarm optimization compared to the original versions. However, comparing all three algorithms for the same problem condition yields the multiple-scattered local search as the superior algorithm over the other two hybrid algorithms for cumulative time probability distribution and cumulative run length distribution.

Acknowledgements

The author would like to thank Anisa Utami and Dody Haryadi for their contribution in this research.

Author details

Ade Jamal
University of Al-Azhar Indonesia, Jakarta, Indonesia

*Address all correspondence to: adja@uai.ac.id

References

[1] Goh SL, Kendall G, Sabar NR. Improved local search approaches to solve the post enrolment course timetabling problem. European Journal of Operational Research. 2017;**261**(1):17-29. DOI: 10.1016/j.ejor.2017.01.040

[2] Elmohamed MAS, Fox G, Coddington P. A comparison of annealing techniques for academic course scheduling. DHPC-045, SCSS-777; 1998

[3] Myszkowski P, Norbeciak M. Evolutionary algorithms for timetable problems. Annales UMCS, Informatica. 2003;**1**(1):115-125. Available from: http://www.annales.umc.lublin.pl

[4] Phillips AE, Walker CG, Ehrgott M, Ryan, DM. Integer programming for minimal perturbation problems in university course timetabling. In: Proceeding of 10th International Conference of the Practice and Theory of Automated Timetabling (PATAT 2014); August 2014; York, United Kingdom; 2014. pp. 26-29

[5] Al-Betar MA, Abdul Khader AT. A harmony search algorithm for university course timetabling. Annals of Operations Research. 2012;**194**(1):3-31. DOI: 10.1007/s10479-010-0769-z

[6] Moody D, Kendall G, Bar-Noy A. Constructing initial neighborhoods to identify critical constraints. In: Burke EK, Gendreau M, editors. Proceedings of the 7th International Conference on the Practice and Theory of Automated Timetabling (PATAT '08); August 2008; Montréal, Canada; 2008

[7] Lewis R, Paechter B. Application of the grouping genetic algorithm to university course timetabling. In: Raidl G, Gottlieb J, editors. Evolutionary Computation in Combinatorial Optimization. Berlin, Germany: Springer; 2005. pp. 144-153. LNCS 3448

[8] Massoodian S, Esteki A. A hybrid genetic algorithm for curriculum based course timetabling. In: Burke EK, Gendreau M, editors. Proceedings of the 7th International Conference on the Practice and Theory of Automated Timetabling (PATAT'08); August 2008; Montréal, Canada; 2008

[9] Jamal A. Solving university course scheduling problem using improved hill climbing approach; In: Proceeding of the International Joint Seminar in Engineering; August 2008; Jakarta, Indonesia; 2008

[10] Abramson D. Constructing school timetables using simulated annealing: Parallel and sequential solutions. Management Science. 1991;**37**(1):98-113. DOI: 10.1287/mnsc.37.1.98

[11] Meyers C, Orlin JB. Very large scale neighbourhood search in timetabling problems. In: Proceeding of the 6th International Conference on the Practice and Theory of Automated Timetabling (PATAT '06); Brno, Czech Republic; 2006

[12] Akinwale OC, Olatunde OS, Olusayo OE, Temitayo F. Hybrid metaheuristic of simulated annealing and genetic algorithm for solving examination timetabling problem. International Journal of Computer Science and Engineering - IJCSE. 2014;**3**(5):7-22

[13] Lawal HD, Adeyanju IA, Omidiora EO, Arulogun OT, Omotosho OI. University examination timetabling using Tabu Search. International Journal of Scientific and Engineering Research. 2014;**5**:10. Available from: http://www.ijser.org

[14] Leighton FT. A graph coloring algorithm for large scheduling problems.

Journal of Research - The National Bureau of Standards. 1979;**84**(6):489-506. DOI: 10.6028/jres.084.024

[15] Dandashi A, Al-Mouhamed M. Graph coloring for class scheduling. In: Proceeding of the 8th ACS/ IEEE International Conference on Computer Systems and Applications (AICCSA 2010); Hammamet, Tunisia; May 2010

[16] Soria-Alcaraz JA, Özcan E, Swan J, Kendall G, Carpio M. Iterated local search using an add and delete hyper-heuristic for university course timetabling. Applied Soft Computing. 2016;**40**:581-593. DOI: 10.1016/j. asoc.2015.11.043

[17] Jamal A. University course scheduling using the evolutionary algorithm. In: Proceeding of International Conference on Soft Computing, Intelligent System, and Information System (ICSIIT 2010); Bali, Indonesia; 2010. pp. 86-90

[18] Jamal A. A three stages approach of evolutionary algorithm and local search for solving the had-m and soft constrained course scheduling problem. In: Proceeding of the 11th Seminar on Intelligence Technology and its Application (SITIA2010); Surabaya, Indonesia; 2010. pp. 324-328

[19] Lutuksin T, Pongcharoen P. Experimental design and analysis on parameter investigation and performance comparison of ant algorithms for course timetabling problem. Naresuan University Engineering Journal. 2009;**4**:31-38

[20] Oner A, Ozcan S, Dengi D. Optimization of university course scheduling problem with a hybrid artificial bee colony algorithm. In: Proceeding of 2011 IEEE Congress of Evolutionary Computation (CEC 2011); 2011. pp. 339-346

[21] Bolaji AL, Khader AT, Al-Betara MA, Awadallah MA. University course timetabling using hybridized artificial bee colony with hill climbing optimizer. Journal of Computational Science. 2014;**5**(5):809-818. DOI: 10.1016/j. jocs.2014.04.002

[22] Ojha D, Sahoo RK, Das S. Automatic generation of timetable using firefly algorithm. International Journal of Advanced Research in Computer Science and Software Engineering. 2016;**6**(4):589-593

[23] Poli R, Kennedy J, Blackwell T. Particle swarm optimization: An overview. Swarm Intelligence. 2007;**1**(1):33-57. DOI: 10.1007/ s11721-007-0002-0

[24] Shiau DF. A hybrid particle swarm optimization for a university course scheduling problem with flexible preferences. Expert Systems with Applications. 2011;**38**(1):235-248. DOI: 10.1016/j.eswa.2010.06.051

[25] Jamal A. Multiple local scattered local search for course scheduling problem. In: Proceeding International Conference on Soft Computing, Intelligent System and Information Technology (ICSIIT 2017) IEEE; September 2017; Bali, Indonesia; 2017. DOI: 10.1109/ICSIIT.2017.22

[26] Forrest S, Mitchell M. Relative building-block fitness and the building-block hypothesis. In: Whitley D, editor. Foundations of Genetic Algorithms 2. San Mateo, CA: Morgan Kaufmann; 1993

[27] Trabzon SA, Pehlivan H, Dehkharghani R. Adaptation and use of artificial bee colony algorithm to solve curriculum-based course time-tabling problem. In: Proceeding of the 5th International Conference on Intelligent Systems, Modelling and Simulation; 2014. pp. 77-82

[28] Burke EK, Elliman DG, Weare RF. A genetic algorithm based university timetabling system. In: Proceeding of the 2nd East-West International Conference on Computer Technology in Education; September 1994; Crimea, Ukraine; 1994. pp. 35-40

[29] Hoos HH. Stochastic local search: Methods, models, application [Thesis]. Darmstadt, Germany: Technisen Universitat Darmstadt; 1998

[30] Hoos HH, Stutzle T. Stochastic local search: Foundation and applications. San Francisco: Elsevier; 2004

Chapter 2

Real-Time Scheduling Method for Middleware of Industrial Automation Devices

Hong Seong Park

Abstract

In this study, a real-time scheduling algorithm, which supports periodic and sporadic executions with event handling, is proposed for the middleware of industrial automation devices or controllers, such as industrial robots and programmable logic controllers. When sensors and embedded controllers are included in control loops having different control periods, they should transmit their data periodically to the controllers and actuators; otherwise, fatal failure of the system including the devices could occur. The proposed scheduling algorithm manages modules, namely, the thread type (or .so type) and process type (or .exe type), for periodic execution, sporadic execution, and non-real-time execution. The program structures for the thread-type and process-type modules that can make the proposed algorithm manage the modules efficiently are suggested; then, they are applied in periodic and sporadic executions. For sporadic executions, the occurrences of events are first examined to invoke the execution modules corresponding to the events. The proposed scheduling algorithm is implemented using the Xenomai real-time operating system (OS) and Linux, and it is validated through several examples.

Keywords: real-time scheduler, middleware, automation device, industrial robot, PLC, periodic execution, sporadic execution, thread type, process type

1. Introduction

Currently, there are many studies about Industry 4.0 [1–3], where numerous industrial automation devices such as industrial robots, programmable logic controllers (PLCs), and industrial Internet of things (IIOT) are used. Manufacturing processes in smart factories have achieved increased flexibility through robots, PLCs, and smart devices, which enable the production of various types of products. Because the robots and PLCs used in these factories are working with humans, the safety of human workers is critical and should be guaranteed. Note that PLCs generally control conveyors and the flow of production processes. In particular, sensors and embedded controllers should transmit their data periodically to the controllers and actuators according to the preset control periods if they are included in control loops having different periods. If some data transmissions fail, fatal failure of the system including the devices could occur. Hence, real-time characteristics, namely, periodic and sporadic, are extremely important for those devices. The periodic characteristic is required for operating the manufacturing system stably and safely, whereas the sporadic characteristic is needed to cope with safety issues.

In addition, mobile manipulators, which consist of mobile platforms and industrial robots, can make the production line more flexible. If this flexibility is expanded to some extent, the manufacturing process can become reconfigurable. However, it is required that the production line is also reconfigurable. A conventional production line is based on a long conveyor system controlled by a PLC, which can obstruct the reconfigurability of the production line. Moreover, the PLC is one of the most important automation devices, and its functional specifications including motion controls are standardized [4–5]. Hence, it is currently being implemented in the software (SW) of embedded controllers and used widely in industrial fields such as smart factories and industrial robots.

In general, industrial robots used in factories utilize PLCs because they must be able to move parts from one cell to another or assemble parts in a cell. Thus, if the production line is composed of two or more cells, which are implemented as moving units based on robots and PLCs, the production line can be reconfigurable. This means that the industrial robot systems used in the lines must perform various types of functions such as manipulation and moving of parts. Hence, the controller of industrial robot systems used in lines manages the motion control SW for manipulation and the PLC function for conveyors and grippers. Some current PLC products can simultaneously manage both conveyors and industrial robots but not all types of industrial robots [6, 7]. Industrial robots and PLCs can control both motion SW for manipulation and PLC functions. Furthermore, they should exchange data via communication among servers and various types of sensors because the program and measured data must be transmitted to other automation devices.

Information technology (IT) is at the center of these technologies. It is however difficult to integrate the rapidly developing IT with conventional robot systems and PLCs. Consequently, middleware technology has been studied for automation devices such as robots and PLCs [4–19]. The middleware used in automation devices manages the processes/threads related to manipulations, vision recognition, PLC functions, transmission of various types of data, safety, and security. This article focuses on the management of processes/threads, which is called real scheduling.

There are some middleware that can be used for automation devices [6–21]. Well-known examples are the CORBA [20], OPC-UA [21], the ones used in CoDeSys [6, 7] and TwinCAT [8], ROS [9, 10], OPRoS [11, 12], openRTM [13, 14], and OROCOS [15, 16]. Among these examples, the OPC-UA and ROS are a type of communication middleware. The ROS manages the execution periods of SW modules using the sleep function, but when SW modules are executed as a process type, it is difficult to keep the period of these modules accurate. Hence, most of the real-time operating systems (OSs) utilize the thread type to keep real-time characteristics. The CoDeSys and TwinCAT support a runtime system that executes control SW modules in real time, which is thought of as a type of middleware. Note that control SW modules used in the CoDeSys and TwinCAT are motion modules for manipulation and PLC functions. The ROS, OPRoS, openRTM, and OROCOS are types of middleware used in robot technology. The CORBA is the most famous middleware supporting communication and management of SW modules, but it is difficult to be implemented in automation devices due to the large size of SW.

The ROS is a popular open SW in the robot field and has been mainly implemented on Linux, but the OPRoS, OROCOS, and openRTM are performed on various types of OSs such as Windows, Linux, and real-time OS. The former executes SW modules as process types, whereas the latter executes SW modules as thread types. Note that general users can use the process type with ease but have difficulty in using the thread type because of its debugging issues and special format. However, the real-time characteristics of SW modules are kept more easily in the thread type. Hence, most of the real-time OSs provide only thread types of SW modules for real time. Because the middleware is utilized in various types of OSs including real-time OSs, it should have a real-time scheduler, which can manage the SW modules in real-time whether they are of process type or thread type. Note that the OPRoS, OROCOS, and openRTM support only thread types of SW modules for real-time services.

Real-time services can be classified into periodic services and sporadic services. In particular, sporadic services of an SW module are processed as follows: the middleware checks the occurrence of an event and then executes the SW module related to that event. Hence, it is necessary for the middleware to support the event handling of sporadic services. Middleware systems applied to industrial robot controllers are the OPRoS and OROCOS, but they do not support sporadic services in real time.

Real-time schedulers of middleware are generally designed and developed based on the execution lifecycle (or state machine), which consists of some states such as IDLE, EXECUTING, DESTRUCTED, and ERROR and is described in the next section. Note that the execution lifecycle is applied only to thread-type SW modules but not to process-type SW modules. In other words, the scheduler controls state transitions to enter into the target state and then manage the real-time threads in a safe manner. However, it is difficult for process-type SW modules to be managed by middleware as seen in the ROS. Hence, the OPRoS, openRTM, and OROCOS manage only the thread-type modules. If the real-time scheduler processes the modules as independent threads, the overhead time including context switching can be critical in the case where the shorter period (e.g., 100 μs) is used. The overhead time needs to be reduced. In addition, it is necessary to consider the execution of process-type SW modules so that users can utilize the middleware easily.

Industrial automation devices used in Industry 4.0 should have flexibility, which can be provided by middleware with real-time schedulers and reliable communication. Real-time schedulers play important roles in supporting reliable and real-time communication. Real-time schedulers for industrial automation devices such as industrial robots [22] should have the following properties of P1–P6:

- (P1) support both process and thread types as execution models of SW modules.
- (P2) support periodic services and sporadic services as real-time services.
- (P3) support non-real-time service if necessary.
- (P4) keep jitters within the minimum bound.
- (P5) support the user-defined priority for each SW module.
- (P6) support the configuration of the SW modules that users can set up.

Examples of processes and threads can be motor control modules, multiple robot control modules, kinematic modules, path planning modules, PLC modules, human-robot interaction modules, and object recognition modules. Motor control modules are executed according to different periods, and PLC modules can be performed cyclically or periodically.

This study proposes a real-time scheduler that satisfies the properties listed above. To reduce the overhead time among threads, the proposed scheduler calls directly the related methods (or functions) of modules in the thread type, where the modules are loaded in .so type in Linux. In addition, it checks the event occurrences to process the corresponding SW modules as sporadic services and then invokes the corresponding SW modules if the related event condition is satisfied.

For this purpose, the middleware provides an event handling function. This study implements the proposed scheduler using Xenomai [23]. Some examples are given to validate the proposed scheduler and show that the worst-case jitters in thread/process types of modules are kept within the minimum bound and that the middleware is performed on Xenomai and Linux.

In Section 2, the requirements of real-time schedulers for middleware of industrial automation devices are proposed. The real-time scheduling algorithm and the program structures for periodic and sporadic executions are suggested, where those executions based on the state machine are related to the thread-type modules. In Section 3, some examples are presented to validate the proposed scheduler. Finally, the conclusions drawn are given in Section 4.

2. Real-time scheduler for middleware of industrial automation devices

2.1 Requirements

In industrial automation devices, such as industrial robots and PLCs for process control, most of the SW modules should be executed periodically. Obviously, PLCs used in discrete I/O controls such as control of conveyors are executed cyclically. Moreover, embedded controllers used in automation devices can execute both manipulation control of industrial robots and control of digital I/Os. Because SW modules used in those automation devices may have different execution periods, it is necessary to set the execution periods smoothly according to the target applications. For example, let us consider SW modules A, B, and C in application 1, which are executed at periods of 10, 30, and 20 ms, respectively. The same modules can be executed at periods of 15, 60, and 30 ms in application 2. The period of a module can vary depending on the application even though the same module is used; thus, it is necessary to set the period smoothly. In general, two terms, namely, basic period and macro period, are utilized in periodic applications. The former is computed using the greatest common divisor of the periods of the modules in the given application, whereas the latter is computed using their least common multiple.

The proposed real-time scheduler is designed and implemented to satisfy the following requirements, which are derived from properties P1–P6 mentioned in Section 1:

- Should support periodic services, sporadic services, and non-real-time services.

- Periodic/sporadic services are divided into thread and process types, and the corresponding information should be provided.

- Should support the process types of legacy SW modules, which can be performed in periodic, sporadic, and non-real-time modes.

- Should be triggered by an event so that sporadic services are performed.

- The event condition is enrolled so that the event handling function can be processed. If the enrolled event condition is satisfied, the corresponding sporadic service is invoked, regardless of the type (whether process type or thread type).

- The event handling function is executed periodically to check the event conditions.

- Periodic services have the highest priority and sporadic services the next priority. Periodic modules and sporadic modules can have different priorities as independent modules, regardless of the type.
- Should use a timer-based operation mode to keep jitters of thread and processes within the minimum bound.
- Should utilize a configuration file written in XML so that users can provide information related to the SW module to the middleware.

Particularly note that events considered in this study are not hardware-driven types because the middleware is executed over the OS. The scheduler in the next subsection is proposed based on these requirements.

2.2 Algorithm for real-time scheduler

The information for users to provide to the middleware is listed below:

- Module types (thread or process)
- Service/operation mode (periodic, sporadic, or non-real-time)
- Module name (file name)
- Period (for periodic service) or deadline (for sporadic service)
- Priority (lower priority or higher)
- Property (input parameters needed to execute the file)

The middleware reads the XML file containing this information and processes it accordingly.

An example of a file in which the above information is stored is shown in **Figure 1**, and its file name is module.xml written in XML. Note that the time unit in the file is nanosecond (ns).

Figure 2 shows a brief algorithm of the proposed real-time scheduler. In main() function, the algorithm reads the configuration file named module.xml and builds two tables, i.e., periodic scheduling table and sporadic scheduling table, according to the computed basic period and priorities of modules. After that, the method "scheduler()" and the basic period are set to link to the timer interrupt and then are periodically executed according to the basic period. The method scheduler() manages the periodic and sporadic threads and processes. The periodic and sporadic processes are managed via signals of scheduler(). Non-real-time modules are executed after the thread of scheduler() has linked to the timer interrupt routine. That is, the execution of the non-real-time modules is independent of the proposed real-time scheduler. The scheduler does not manage the non-real-time modules to reduce the computation time.

The middleware reads the configuration file such as module.xml in **Figure 1** and computes the basic period of 100 μs and the macro period of 600 μs. Using these two periods, the middleware generates the periodic scheduling table shown in **Figure 3** and the sporadic scheduling table shown in **Figure 4**. Note that pRun() denotes the pointer of the function run(), which is described in Section 2.3.

```xml
<?xml version="1.0" encoding="UTF-8"?>
<!-- filename: file to be loaded and executed by middleware -->
<!-- moduletype: process(exe type), thread(so or dll type) -->
<!-- operationtype: periodic, sporadic, non-real -->
<!-- period: nano sec -->
<!-- priority: the lowest value is the highest priority -->
<!-- property: input parameters needed to execute the module -->
<root>
  <module> <!-- periodic module with period of 100 us -->
     <filename>build/controller1.so</filename>
     <moduletype>thread</moduletype>
     <operationtype>periodic</operationtype>
     <period>600000</period>
     <priority>3</priority>
     <property>
        <value name="counter">5</value>
     </property>
  </module>
  <module>
     <filename>./control2.so</filename>
     <moduletype>thread</moduletype>
     <operationtype>periodic</operationtype>
     <period>100000</period>
     <priority>2</priority>
  </module>
  <module>
     <filename>./control3.so</filename>
     <moduletype>thread</moduletype>
     <operationtype>periodic</operationtype>
     <period>100000</period>
     <priority>1</priority>
  </module>
  <module>
     <filename>./control4.exe</filename>
     <moduletype>process</moduletype>
     <operationtype>periodic</operationtype>
     <period>300000</period>
     <priority>4</priority>
  </module>
  <module> <!-- sporadic module with deadline of 100 us -->
     <filename>./emergency.so</filename>
     <moduletype>thread</moduletype>
     <operationtype>sporadic</operationtype>
     <deadline>1000000</deadline>
     <priority>1</priority>
  </module>
  <module>
     <filename>./vision.exe</filename>
     <moduletype>process</moduletype>
     <operationtype>sporadic</operationtype>
     <deadline>100000000</deadline>
     <priority>2</priority>
  </module>
  <module> <!-- non-real-time module -->
     <filename>./monitoring.exe</filename>
     <moduletype>process</moduletype>
     <operationtype>non-real</operationtype>
  </module>
</root>
```

Figure 1.
Module.xml file.

```
main()
{
  read configuration file
  calculate the basic period and the macro period
  load the thread-type SW modules and the process-type SW modules
  set the periodic scheduling table and sporadic scheduling table.
  Invoke start() methods of the periodic modules and sporadic modules
  set basic period and the thread, scheduler(), to timer interrupt
  execute the non-real-time modules defined in the configuration file
  set this process to the lowest priority
  wait(); // wait for all processes and threads to stop
}
// execute the scheduler in thread mode and manage the periodic and
// sporadic modules.

scheduler()
{
  get the entry from one row of the periodic scheduling table
  while(entry != NULL) {
    // the corresponding method is called if thread type
    if the entry is .so type,
     call the function pRun() of the entry
    else // process type
  send signal to the process to continue its execution
   get the next entry from the row
    }
    increase the row by one to move to the next row of the periodic scheduling table
    get the entry from the sporadic table
     while(entry != NULL)
    {
      If (the entry Is .so type) {
         if (condition() of the entry is satisfied)
           call the function pRun() of the entry
      }
      else
      send signal to the sporadic process to continue its execution
    }
  } // while
}
```

Figure 2.
Brief algorithm for the proposed real-time scheduler.

In **Figure 3**, the index is the execution order. That is, four periodic modules are executed in the first period (index 0) starting at 0 μs. In the first period, control3.so is first executed, and control4.exe is executed last according to the priority in **Figure 1**. Control1.so is executed once every 600 μs, and control4.exe is executed once every 300 μs. Obviously, the real-time scheduler distinguishes threads and processes and executes them properly. The sporadic scheduling table is shown in **Figure 4**. The sporadic modules are listed in order of priority in the table. For execution of the SW modules, the function pointers and the process IDs are stored according to the type (thread or process) in the scheduling table.

index	name of module (its symbol) - ordered by priority
0	control3.so(c3), control2.so(c2), controller1.so(c1), control4.exe(c4)
1	control3.so, control2.so,
2	control3.so, control2.so,
3	control3.so, control2.so,control4.exe
4	control3.so, control2.so,
5	control3.so, control2.so,

repeat

Figure 3.
Example of periodic scheduling table based on ***Figure 1****.*

index	name of module (priority order)
0	emergency.so, vision.exe

Figure 4.
Example of sporadic scheduling table based on ***Figure 1****.*

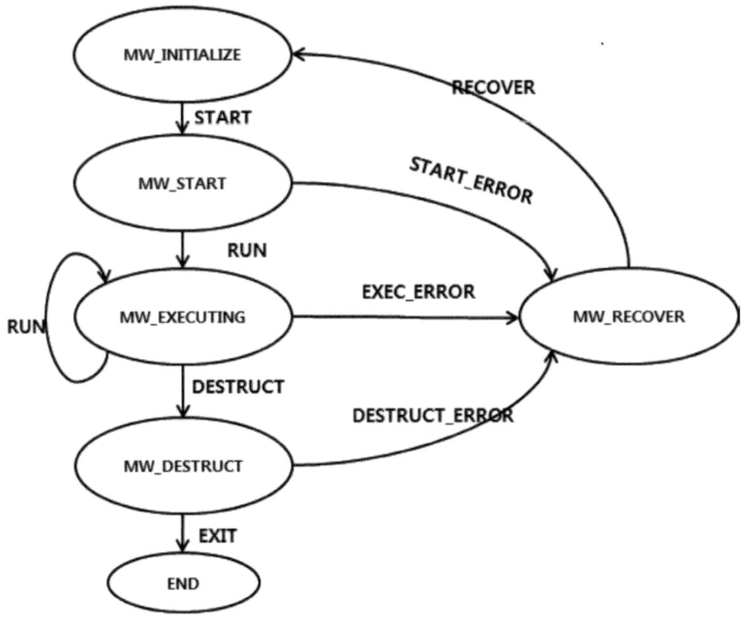

Figure 5.
Execution lifecycle of thread-type SW module.

The thread modules are executed according to the execution lifecycle shown in **Figure 5**. After loading the thread-type SW modules, the module is initialized by the method initialize(), which is illustarted in **Figure 8**, and the module enters into the MW_INITIALIZE state. If the method start() is invoked, the module enters into the MW_START state. After all the thread-type modules enter into the MW_START states, execution of the real-time scheduling algorithm, scheduler() is started by invoking the method run(), which is also shown in **Figure 8**. The module is periodically called or receives signal at the MW_EXECUTING state. After completion of the module execution, the module invokes the method destroy() and then enters into the MW_DESTRUCT state, and consequently, the execution of the module is ended. In the MW_RECOVER state, some error handling is possible using recover-related methods. Note that the scheduler directly calls or invokes the method run() of modules to reduce the OS overhead time such as context switching time.

Process-type modules are also executed, and they immediately wait for the period-starting signal. If a module receives signals from the scheduler, the module is re-executed from the waiting position.

The operation of the scheduler in **Figure 2** is shown in **Figure 6**. In this figure, it can be observed that the non-real-time modules can be executed only when a sufficient idle time remains in one period. The scheduler calls the run() method of the modules to reduce the overhead of controlling the threads or processes, where the run() method is shown in **Figures 7** and **8**. After the executions of periodic modules are finished in a period, the scheduler checks the event conditions of sporadic modules. If the condition is true, the scheduler calls the corresponding sporadic method for thread types and sends the signal to the sporadic module for process

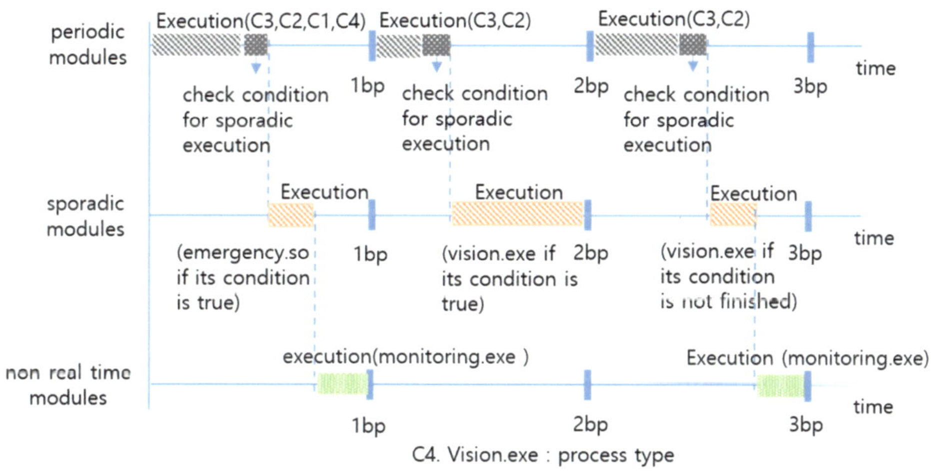

Figure 6.
Operation of real-time scheduler for periodic, sporadic, and non-real-time services based on ***Figures 3*** *and* ***4***.

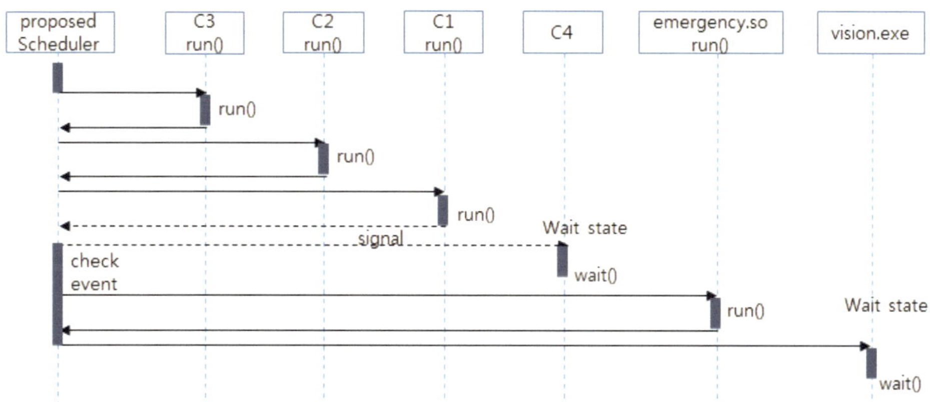

Figure 7.
Example of control scheme of real-time scheduler for periodic and sporadic modules.

```
extern "C"{
 void initialize(irp::Property const& property){
     ... //user program for initialize() method
 }
 void start(void){
             ... //user program for start() method
 }
 void run(void){
             ... //user program for run() method
 }
 void error(int type){
             ... //user program for error() method
 }
 void destruct(void){
             ... //user program for destruct() method
 }
 void recover(void) {
             ... //user program for recover() method
 }
}
```

Figure 8.
Program structure of thread-type periodic module.

type. Note that the modules are executed over the multicore CPU and modules with thread types and process types are executed on different cores. Hence, two types of modules can be executed at the same time.

Periodic and sporadic operations can be divided into thread type and process type as shown in **Figure 6**. The scheduler manages the SW modules according to their types. **Figure** 7 shows an example of a control scheme of a real-time scheduler for periodic and sporadic modules, where the thread-type modules are directly called by the scheduler and the process-type modules are executed by the signal from the scheduler. As shown in **Figure** 7, the periodic modules are executed first. The scheduler executes a module by calling the run() method of the corresponding module in the .so file, where the module has a thread type. To execute a process-type module, the scheduler sends a signal to the process of the module. Note that process-type modules are executed independently of the scheduler as a type of process and the scheduler is also a type of process.

The program structures of thread-type and process-type modules are described in the next subsection.

2.3 Program structures for thread-type and process-type modules

Because the operation method of a thread-type periodic module is different from that of a process type, the program structure of the thread-type periodic module should be different from that of the process type, which are shown in **Figures 8** and **9** respectively. The method initialize() is executed immediately after the module is loaded in the memory. The methods start(), run(), destruct(), error(), and recover() are called

```
int main(int argc, char argv[]){
 initPeriodExe();
        ... //user program for initialize
   while(1){
        waitPeriod();
        ... //user periodic body
   }
   return 0;
}
```

Figure 9.
Program structure of process-type periodic module.

```
extern "C'{
  int condition(){
    check condition;
    if(condition)
          return TRUE; // condition is met.
   else
          return FALSE; //otherwise
  }

  void initialize(void){

 ... //user event initialize
  }
  void start(void){
             ... //user program for start() method
  }
  void run(void){
     ... //user event body
  }
   // similar methods to the .so-type periodic module
   // destruct(), error(), recover() are added
}
```

Figure 10.
Program structure of .so-type sporadic module.

when events such as START, RUN, DESTRUCT, XXX_ERROR, and RECOVER occur, respectively, which are shown in **Figure 5**.

Users should insert proper codes into parts named user program. In general, a legacy program has a structure of a process type, which is simpler than that of a thread type. To use a legacy program on the proposed middleware, two functions

```
int condition() {
   waitSporadic(); // wait for signal
   check condition;
   if(condition)
         return TRUE; // condition is met.
   else
         return FALSE; //otherwise
}
int main(int argc, char argv[]){
   initSporadicExe();
       ... //user program for initialize
   while(1){
       if (!condition()) // if not satisfied
           continue;   // enter into wating state
       ... //user sporadic body
   }
   retrun 0;
}
```

Figure 11.
Program structure of process-type sporadic module.

initPeriodExe() and waitPeriod() should be added in it. initPeriodExe() is a function for enrollment of the corresponding process to the scheduler, and waitPeriod() is a type of function to wait for a signal from the scheduler. Note that the scheduler sends a signal to a process when the corresponding process wants to be executed. Upon receiving the periodic signal, the module transitions from the waiting state to the executing state and then executes the main body, which is the part marked user periodic body in **Figure 9**. The module enters into the waiting state by the waitPeriod() function.

After the execution of periodic modules, the scheduler checks whether any events for sporadic modules have occurred. The modules corresponding to such events are listed in order of priority using EDF (earliest deadline first) method [6–9]. As shown in **Figures 2** and **6**, the scheduler checks periodically by calling or invoking the condition() function in **Figures 10** and **11**. If condition() returns a value of TRUE in **Figure 10**, the scheduler executes the corresponding .so-type module. The process-type sporadic module is designed so that it receives sporadic signals from the scheduler, checks its condition, and executes the user execution body if condition() is satisfied.

3. Evaluation

Experiments were performed using the test cases in **Table 1** on a PC with the following specifications to validate the proposed scheduling algorithm:

- Ubuntu 14.04 LTS (64 bit), kernel: Linux 4.1.18
- Xenomai 3.0.3, ipipe-core-4.1.18
- CPU: Intel(R) Core(TM) i7–7700 CPU @ 3.60 GHz, four cores

The purpose of the experiments was to determine how the periodic modules are affected by the execution of all types of modules. Hence, the worst-case jitter and related jitter statistics are measured and analyzed. Let J_n, T_n, and P_n denote the nth jitter, the starting time of the n-th execution of the target, and the starting time of the nth period, respectively. The jitter considered in this article is calculated as follows:

$$J_n = P_n - T_n = T_0 + \mathrm{n} * \mathrm{period} - T_n, \tag{1}$$

where T_0 denotes the reference time of the target module and period, period denotes a basic period, and $P_n = T_0 + \mathrm{n} * \mathrm{period}$.

The modules were executed using a basic period of 100 μs. In **Table 1**, the type of measured module indicates whether the jitter was measured in a .so (thread) or .exe (process) module. The test results are also presented in **Table 1** and **Figures 12–19**. Note that the jitter is computed using Eq. (1) and it is measured in a special module of periodic modules and the type of the measured module is given in **Table 1**.

Table 1 and **Figures 12–19** show that the ranges of mean values and variances of the process-type periodic module in the test cases are -5.5519 μs to -5.991 μs and 11.598–12.438 μs, respectively. The ranges of mean values and variances of the thread-type periodic module in the test cases are -0.027 μs to -0.105 μs and 0.716–3.772 μs, respectively. The worst-case jitter is 12.438 μs, which is measured in the process-type periodic module, and the jitter rate is 12.438% with respect to the basic period of 100 μs. The worst-case jitter measured in the thread-type periodic

No.	Test cases						Test results		
	Number of periodic modules		Number of sporadic modules		Number of non-real-time modules	Type of measured module	Jitter mean (μs)	Jitter variance (μs)	Worst-case jitter (μs)
	.so	.exe	.so	.exe					
1	1	0	0	0	0	Thread	−0.027	0.000839	3.772
2	15	0	0	0	0	Thread	0.127	0.002037	2.933
3	5	3	0	0	0	Thread	−0.042	0.001263	2.793
4	5	3	0	0	0	Process	−5.723	19.480341	11.598
5	5	3	11	0	0	Thread	0.033	0.001174	0.716
6	5	3	11	0	0	Process	−5.580	19.624376	11.614
7	5	3	5	6	0	Thread	−0.105	0.001633	1.405
8	5	3	5	6	0	Process	−5.991	21.136330	12.438
9	5	3	5	5	8	Thread	0.007	0.001424	1.104
10	5	3	5	5	8	Process	−5.519	20.617964	11.758

Table 1.
Test cases for evaluation of scheduling algorithm.

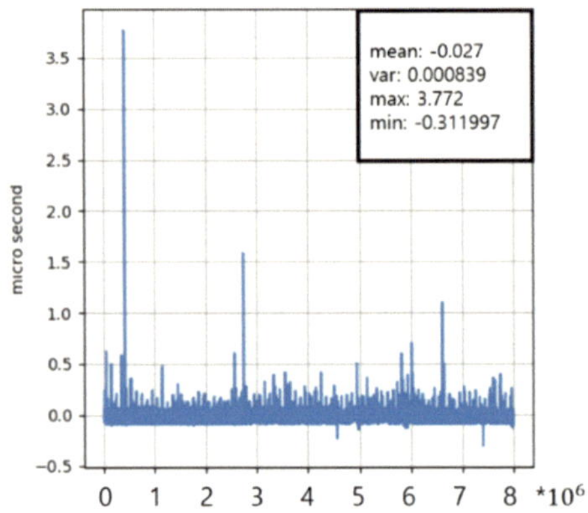

Figure 12.
Jitters in test case 1.

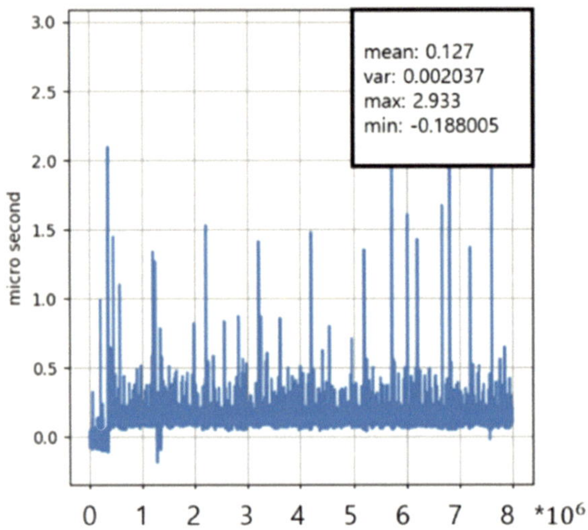

Figure 13.
Jitters in test case 2.

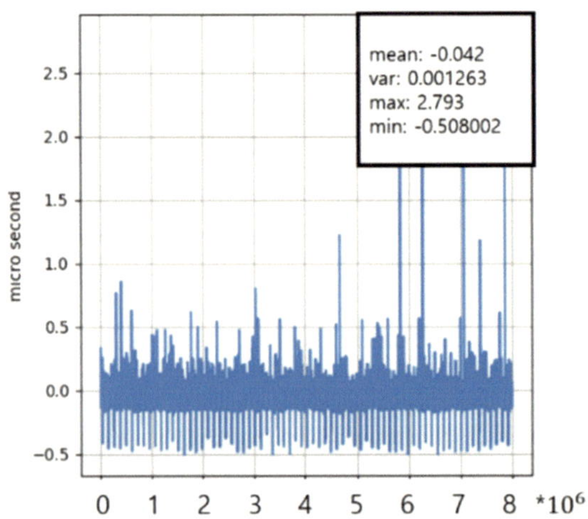

Figure 14.
Jitters in test case 3.

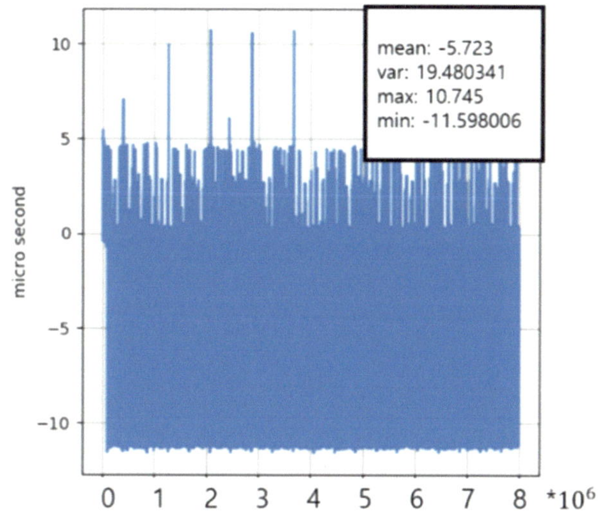

Figure 15.
Jitters in test case 4.

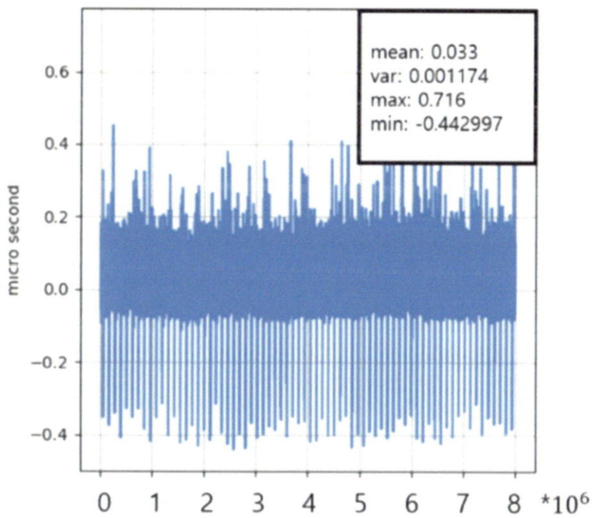

Figure 16.
Jitters in test case 5.

module is 2.933 μs, which is the result of test case 2, and the jitter rate is 2.933% with respect to 100 μs.

For the thread-type periodic module, as the load increases, the variation amount of jitters also increases; however, the worst-case jitter does not exceed 4 μs, and the change in the jitter variances is not significant. For the process-type periodic module, as the load increases, the variation amount of jitters increases significantly; however, the worst-case jitter does not exceed 12.5 μs, and the changes in the jitter variances and worst-case jitters are not large. The test results in **Table 1** and **Figures 11–18** indicate that the proposed scheduling algorithm can be efficiently used by industrial automation devices even for various types of applications.

It is evident from the results in **Table 1** and **Figures 12–19** that the basic period is maintained very satisfactorily in all test cases. This means that the proposed

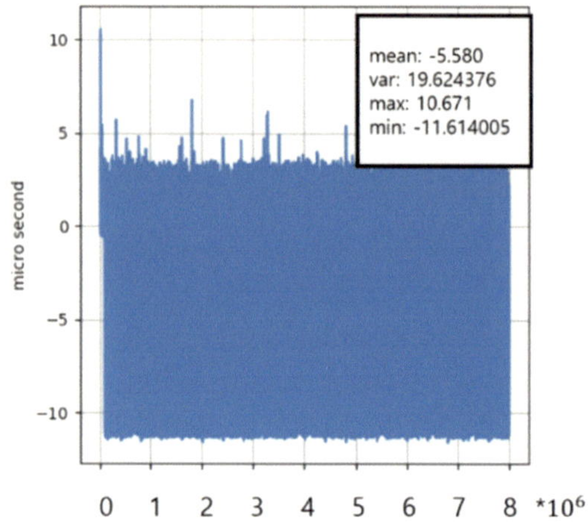

Figure 17.
Jitters in test case 6.

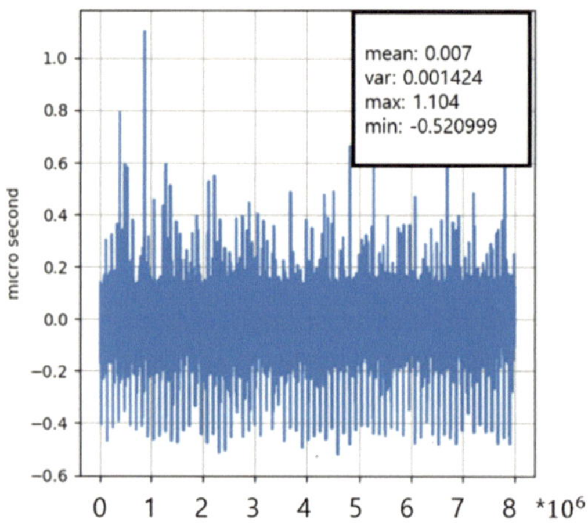

Figure 18.
Jitters in test case 9.

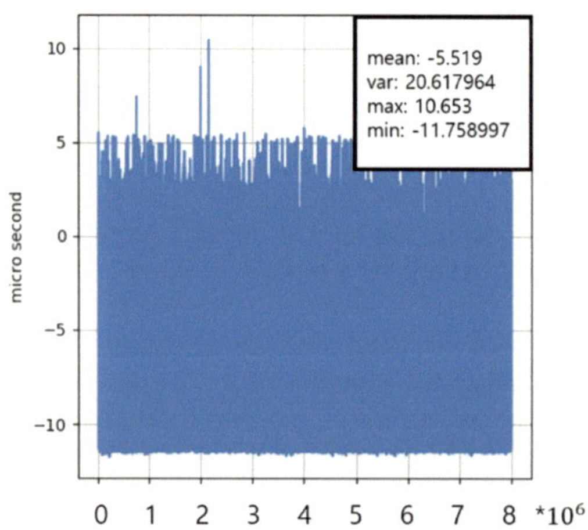

Figure 19.
Jitters in test case 10.

scheduler can work effectively in various situations. Moreover, it can be observed from the test results that the process-type periodic module has a greater effect on jitters than the thread-type and the sporadic modules. Hence, it is better to use the thread-type than the process-type for periodic modules. If the process-type modules as legacy modules are utilized, it is necessary to reduce their number.

4. Conclusion

This study proposed a real-time scheduling algorithm for middleware of industrial automation devices or controllers such as industrial robots and PLCs. The proposed algorithm strictly maintains the periods and supports both periodic and sporadic executions with event handling. It has managed modules, namely, the thread type (or .so type) and process type (or .exe type), for periodic execution, sporadic execution, and non-real-time execution. This study provided the program structures of the thread-type and process-type modules for periodic and sporadic services to manage them efficiently. For sporadic services, the scheduler checks for the occurrence of events using the condition() method in the sporadic modules before invoking the corresponding module.

The proposed scheduling algorithm was implemented using the Xenomai real-time OS and Linux, and it was validated through some test cases. The worst-case jitters measured in the thread-type periodic module and the process-type periodic module were 2.933 and 12.438 μs, respectively, where the jitter rates were 2.933 and 12.438% with respect to the basic period of 100 μs. The basic period was maintained very satisfactorily without missing any periods in all the test cases. The test results showed that the proposed scheduler could work well in various situations. Furthermore, it is better to use the thread-type module than the process-type module when periodic modules are used. It was demonstrated that the proposed scheduling algorithm could be used for the middleware of industrial automation devices or controllers.

In future research, the proposed scheduling algorithm will be tested to handle periodic modules, sporadic events, and non-real-time modules in multicore systems, manage process-type periodic modules with smaller worst-case jitters, and support various types of OSs.

Acknowledgements

This work was partly supported by Korea Evaluation Institute of Industrial Technology (KEIT) grant funded by the Korea government (MOTIE) (No. 10067414, development of real-time-assisting SW platform for industrial robot).

Author details

Hong Seong Park
Department of Electrical and Electronic Engineering, Kangwon National University, South Korea

*Address all correspondence to: hspark@kangwon.ac.kr

References

[1] Rueßmann M, Lorenz M, Gerbert P, Waldner M, Justus J, Engel P, Harnisch M. Industry 4.0: The Future of Productivity and Growth in Manufacturing Industries. Boston Consulting Group; 2015. Available from: https://www.bcg.com/publications/2015/engineered_products_project_business_industry_4_future_productivity_growth_manufacturing_industries.aspx

[2] Thoben K, Wiesner S, Wuest T. Industrie 4.0 and smart manufacturing—A review of research issues and application examples. International Journal of Automation Technology. 2017;**11**:4-16. DOI: 10.20965/ijat.2017.p0004

[3] MacDougall W. Industrie 4.0 smart manufacturing for the future. Available from: https://www.manufacturing-policy.eng.cam.ac.uk/documents-folder/policies/germany-industrie-4-0-smart-manufacturing-for-the-future-gtai/view

[4] IEC. IEC 61131-3 Programmable controllers—Part 3: Programming languages; 2013

[5] PLCopen Technical Committee 2. Function Blocks for Motion Control: Part 3—User Guidelines. 2013. Available from: https://www.plcopen.org/system/files/downloads/plcopen_motion_control_part_3_version_2.0.pdf

[6] CODESYS Group. WHY CODESYS? [Internet]. Available from: https://www.codesys.com/the-system/why-codesys.html

[7] Korobiichuk I, Dobrzhansky O, Kachniarz M. Remote control of nonlinear motion for mechatronic machine by means of CoDeSys compatible industrial controller. Tehnički Vjesnik. 2017;**24**:1661-1667. DOI: 10.17559/TV-20151110164217

[8] Beckhoff. TWINCAT-PLC and motion control on the PC [Internet]. Available from: https://www.beckhoff.com/twincat/

[9] OSRF Site [Online]. Available from: www.ros.org

[10] Wei H, Shao Z, Huang Z, Chend R, Guanb Y, Tanc J, et al. RT-ROS: A real-time ROS architecture on multi-core processors. Future Generation Computer Systems. 2016;**56**:171-178. DOI: 10.1016/j.future.2015.05.008

[11] Han S, Kim M, Park HS. Open software platform for robotic services. IEEE Transactions on Automation Science and Engineering. 2012;**9**:467-481. DOI: 10.1109/TASE.2012.2193568

[12] OPRoS Site [Online]. Available from: www.ropros.org

[13] OpenRTM Site [Online]. Available from: www.openrtm.org

[14] Hasegawa R, Yawata N, Ando N, Nishio N, Azumi T. Embedded component-based framework for robot technology middleware. Journal of Information Processing. 2017;**25**:811-819. DOI: 10.2197/ipsjjip.25.811

[15] OROCOS site [Online]. Available from: www.orocos.org

[16] Rastogi N, Dutta P, Krishna V, Gotewa KK. Implementation of an OROCOS based real-time equipment controller for remote maintenance of tokamaks. In: Proceedings of the Advances in Robotics; June 28-02 July 2017; New Delhi, India; DOI: 10.1145/3132446.3134900

[17] Muratore L, Laurenzi A, Hoffman EM, Rocchi A, Caldwell DG, Tsagarakis NG. XBotCore: A real-time cross-robot software platform. In: 2017 First IEEE

International Conference on Robotic Computing (IRC); 10-12 April 2017; Taichung, Taiwan; DOI: 10.1109/IRC.2017.45

[18] YARP site [Online]. Available from: www.yarp.it

[19] Paikan A, Pattacini U, Domenichelli D. A best-effort approach for run-time channel prioritization in real-time robotic application. In: 2015 IEEE/RSJ International Conference on Intelligent Robots and Systems (IROS); 28 Sept.-2 Oct. 2015; Hamburg, Germany. 2015. pp. 1799-1805. DOI: 10.1109/IROS.2015.7353611

[20] Levine DL, Schmidt DC, Flores-Gaitan S. CORBA measuring OS support for real-time CORBA ORBs. In: 1999 Proceedings Fourth International Workshop on Object-Oriented Real-Time Dependable Systems; 27-29 Jan. 1999; Santa Barbara, CA, USA. 1999. pp. 9-17. DOI: 10.1109/WORDS.1999.806555

[21] OPC Foundation. PLCopen and OPC Foundation: OPC UA Information Model for IEC 61131-3. 2010

[22] Yu D, Park HS. Real-time middleware with periodic service for industrial robot. In: 2017 14th International Conference on Ubiquitous Robots and Ambient Intelligence (URAI); 28 June-1 July 2017; Jeju, South Korea. 2017. pp. 879-881. DOI: 10.1109/URAI.2017.7992853

[23] Xenomai site [online]. Available from: xenomai.org

Section 2

On Addressing Scheduling for Parallel and High-Performance Computing Environments

Chapter 3

Intelligent Workload Scheduling in Distributed Computing Environment for Balance between Energy Efficiency and Performance

Larysa Globa, Oleksandr Stryzhak, Nataliia Gvozdetska and Volodymyr Prokopets

Abstract

Global digital transformation requires more productive large-scale distributed systems. Such systems should meet lots of requirements, such as high availability, low latency and reliability. However, new challenges become more and more important nowadays. One of them is energy efficiency of large-scale computing systems. Many service providers prefer to use cheap commodity servers in their distributed infrastructure, which makes the problem of energy efficiency even harder because of hardware inhomogeneity. In this chapter an approach to finding balance between performance and energy efficiency requirements within inhomogeneous distributed computing environment is proposed. The main idea of the proposed approach is to use each node's individual energy consumption models in order to generate distributed system scaling patterns based on the statistical daily workload and then adjust these patterns to match the current workload while using energy-aware Power Consumption and Performance Balance (PCPB) scheduling algorithm. An approach is tested using Matlab modeling. As a result of applying the proposed approach, large-scale distributed computing systems save energy while maintaining a fairly high level of performance and meeting the requirements of the service-level agreement (SLA).

Keywords: energy efficiency, performance, SLA, distributed computing system, scheduling, horizontal scaling

1. Introduction

Nowadays information technologies penetrate all spheres of human life. According to the Gartner Top 10 Strategic Technology Trends for 2019 [1], the new world-driving trends are going to include augmented analytics, immersive technologies, edge computing and blockchain in the nearest future. These technologies require extremely highly efficient distributed computing systems that could process large amounts of data, consuming as little resources as possible.

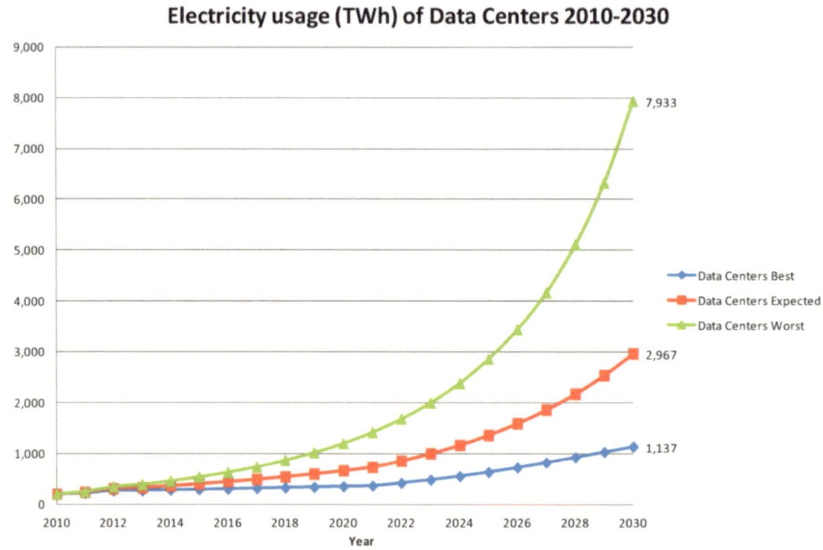

Figure 1.
Electricity usage of data centres 2010–2030 estimation [2].

Electrical power is one of the most demanded resources for the large-scale distributed computing systems. Both science and industry have made tremendous efforts to reduce the power consumption of large-scale computing systems over the past 10 years. According to the Huawei Technologies Sweden estimation, presented in [2], power consumption of data centres all over the world is still going to grow in the nearest future. However, due to the newly developed techniques, we can expect much lower growth up to 2967 TWh. On the contrary, in the worst case, data centres could consume up to 7933 TWh of energy without the use of any energy-saving approaches (**Figure 1**).

Although the fruitful cooperation between science and industry has already brought very good results in terms of reducing data centre power consumption, there is still a very high demand for approaches that would allow to cope with new challenges associated with the rapid development of distributed computing.

Within this study we propose an intelligent workload scheduling approach, which is aimed at improving energy efficiency of distributed computing systems through the application of energy-aware scheduling combined with scaling, taking into account data processing performance as well. An approach considers first of all inhomogeneous distributed systems designed to use cheap commodity hardware as much as possible.

The chapter is structured as follows: Section 2 contains state-of-the-art analysis of distributed computing system energy efficiency problem. Section 3 explains the problem to be solved by proposed approach. Section 4 introduces proposed intelligent workload scheduling approach efficiently combined with dynamic scaling approaches. Section 4.1 presents a model of proposed approach implementation, and Section 5 concludes the work with a summary and outlook on future work.

2. State of the art and background

Many approaches to increasing energy efficiency of the large-scale distributed computing systems (power management approaches) already exist. According to the paper [3], these approaches can be divided into static and dynamic in terms of decision-making process. Both static and dynamic approaches can be applied either at the hardware level or at the software level. At the same time, these approaches are also classified according to the scope (cloud environment, single server, etc.). The overall structure of approach classification proposed in [3] is depicted in **Figure 2**.

Since the load on computing systems usually changes dynamically (in particular, in the cloud environment), it is worth to pay attention to the dynamic approaches to power management, respectively. Thus, within this study we mainly focus on dynamic energy-efficient approaches.

Considering hardware-based approaches, one of the most common approaches is dynamic voltage and frequency scaling (DVFS). DVFS is a power management technique that is effective in reducing power dissipation by lowering the supply voltage [4]. DVFS is widely used to manage the energy and power consumption in modern processors; however, for DVFS to be effective, there is a need to accurately predict the performance impact of scaling a processor's voltage and frequency [5, 6]. Moreover, the implementation of hardware-based approaches may be challenging, especially when it comes to inhomogeneous systems consisting of commodity hardware. Such approaches usually require additional expenses in order to adapt the system to their use. However, hardware-based approaches are very efficient. And they might bring even more energy efficiency when used in conjunction with software-based techniques.

Among the software approaches, the most commonly used are scheduling and consolidation [7]. Scheduling approaches are aimed at distributing the workload among the servers in a way that no servers are underutilized, jobs' processing performance is high enough and, in the case of energy-aware scheduling, the power consumption of the whole computing system is minimized. Consolidation approaches concern virtualized environments and are designed to balance virtual machines (VMs) so that they can run on as few servers as possible. Idling servers are then shut down or switched to a standby mode. It means that the consolidation is tightly coupled with horizontal scaling approaches.

The authors of [8] proposed the performance and energy-based cost prediction framework that dynamically supports VMs auto-scaling decision and demonstrates the trade-off between cost, power consumption and performance. This framework allows to estimate auto-scaling total cost, which is essential when using consolidation and horizontal auto-scaling approaches. Dynamic auto-scaling can be a big gain in energy efficiency, but estimating the cost of automatic scaling should not lead to excessive overhead. Thus, in some cases, it may be worthwhile to avoid constant dynamic auto-scaling and basically rely on statistics instead, while adjusting the scale of the system when it is really needed.

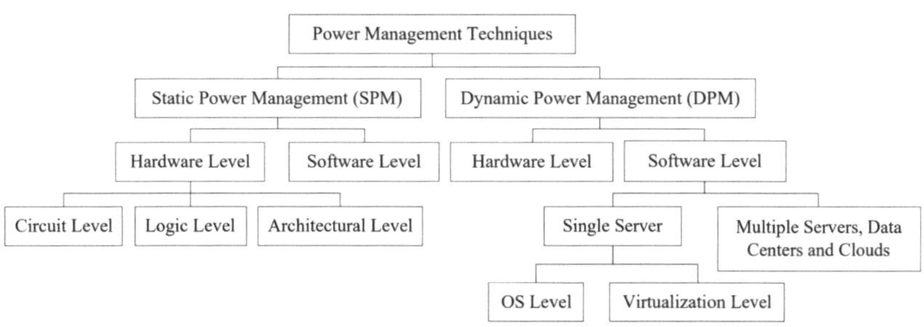

Figure 2.
Classification of power management approaches to increasing energy efficiency of computing systems [3].

In this chapter we use this idea, and we are mainly focused on the efficient combination of horizontal scaling and energy-aware scheduling approaches. In the field of energy-aware scheduling, there are many approaches developed so far. Some of them are also aimed to combine different fundamental ideas of energy-efficient computing.

The authors of the paper [9] proposed adaptive energy-efficient scheduling (AES) approach which combines scheduling with the dynamic voltage scaling (DVS) technique. This approach is proposed to be used in homogeneous computing systems. It consists of two phases—in the first phase, an adaptive threshold-based task duplication strategy is used, which can adjust the optimal threshold according to the specified schedule length, network power, processor power and application; in the second phase, DVS-enabled processors that can scale down their voltages are used for processing. A proposed approach gives a gain of 31.7% in terms of energy efficiency without performance loss that is quite a good result. Within this study we also use an idea of combining scheduling approaches with other methodologies; however, the authors of [9] use DVS and are limited by using DVS-enabled hardware. Their approach is designed for homogeneous computer environments, while we are focused on inhomogeneous environments, since it makes sense for service providers to use cheap commodity hardware with different physical parameters in order to reduce capital expenses.

Another example of energy-aware scheduling approach is Min_c [10]. This strategy takes into account the variety of tasks that comes to the computing system and the fact that the resources required by these tasks are different. The main drawback of this approach is that the model of energy consumption is the same for each node. It has nonlinear character that is close to reality, but the characteristics of different machines can differ. Therefore, we propose defining $P = f(CPU)$ dependencies for each machine individually.

Within our previous research [11–13], energy-aware scheduling approach called power consumption and performance balance (PCPB) was proposed and experimentally tested. In [13] two possible modifications to the PCPB were described—one of them uses tasks classification; another one applies scaling in addition to energy-aware scheduling in order to further increase energy efficiency. It was determined that developed energy-aware scheduling approach PCPB gives the best results in terms of energy efficiency (energy savings up to 33.59%) while being used in conjunction with horizontal scaling (scale-in and scale-out). Thus, in this chapter we elaborate on this basic idea and propose to enhance PCPB with scaling techniques and smartly power off and on idling servers while distributing the workload between them using PCPB.

3. Problem definition

Consider a distributed computing system consisting of N nodes. Each node N_j is described by the following:

- V_j—the amount of available RAM
- $flops_j$—the productivity of the node N_j, which has k_cores_j of the computing cores
- $P_j = f(CPU_j)$—the power consumption function of the node, which is experimentally determined for each N_j

Job is the unit of computing. The component of job is called a task. Physically, one job is represented as a single computing process of a computer. Subprocesses or child processes in turn appear to be tasks.

We will use the term “job” to determine the atomic computational problem that can be located and executed on one of the computing nodes of the system (i.e. “job” is the unit of load distribution). The income jobs form the queue.

Let us introduce the assumptions:

1. The queue consists of Q positions.

2. Jobs come to the queue at random moments of time τ. If the length of the queue is $0 < l < Q$, the job is placed on a free space in the queue. Otherwise, the job is rejected.

3. All jobs in the queue are independent to each other.

4. The job may have one of the five possible states: “preparation for execution”, “readiness for execution”, “in process of execution”, “execution successfully completed”, and “execution interrupted”.

5. Each ith job is characterized by the following parameters:

    - The volume of job—the number of floating-point operations that must be performed by the machine within this job (the number of floating-point operations is not natively used to measure the jobs’ volume; however, we use this artificial unit to express the amount of work that is required by job to be performed):

$V_i = const\ [operations]$

    - The maximum execution time (or timeout):

$\Delta t_{maxi} = const\ [\sec]$

is the time period from the moment the job arrives at the processing, after which it must be completed. If the job has not been transferred to the state of “execution successfully completed” after the time Δt_{maxi}, it is transferred to the state of “execution interrupted” and is output from the system.

    - Minimal amount of resources required to complete the job:

        - Minimal amount of RAM needed:

$RAM\_\min\_i\ [Gb]$

        - Minimal required number of processing cores:

$k\_cores\_\min\_i\ [cores]$

An optimal parallelization for the task is not considered within this study. Let us consider that the tasks are parallelized beforehand and just declared how many cores they require to be processed, as an input data for processing system.

        - Minimal required volume of persistent storage:

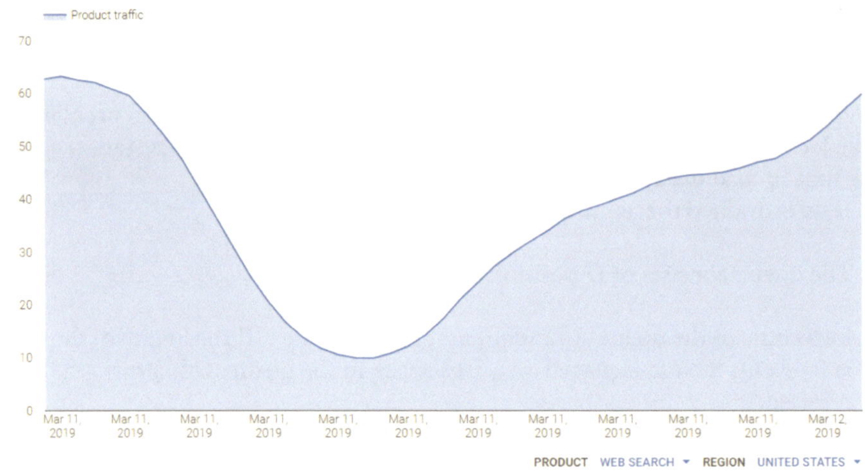

Figure 3.
Workload statistic for the Google web search service (11 March 2019) [14].

*storage*_min_*i* [*Gb*]

- Job's priority is an optional numeric parameter that can be set for a job and determine the priority of its execution in comparison with other jobs. Let the priority $priority_i$ be determined by an integer from 1 to 10. In this case, priority processing requires tasks with priority 1. Let the default value of the priority of the job be 10 units.

Consider the daily workload curve to be a part of an input data for the problem. Every service provider is able to gather the daily statistics of the workload in his computing system. As a result, the pattern of the day workload can be generated. For example, Google provides such statistics of using its web search service in the Transparency Report [14] (**Figure 3**).

Under the certain circumstances (holidays, festivals, special events, etc.), this pattern can be changed. It means that it is possible to create certain system configuration patterns for the stable workload, respectively, but there is a need to adapt the system in the case of changes being foreseen.

Given an input data, the problem is formulated as follows: to reduce total power consumption of considered computing system while increasing the performance of data processing and meeting service-level agreement (SLA) requirements.

4. Proposed approach

Within the previous studies [11–13], it was determined that the use of energy-aware scheduling is most effective when used in conjunction with consolidation approaches and further scaling the whole system in and out. In [13] it was shown that powering off idle servers could save up to 33.59% of the energy in the considered computing cluster.

Within the current study, the system model and main idea of proposed approach were modified and can be described as follows:

1. To define experimentally individual dependencies $P_j = f(CPU_j)$ for all computing nodes of the system as power consumption functions. Having power consumption functions and performance value of each node, to range all nodes according to their integrated performance and power consumption criteria from the best one to the worst one.

2. To use energy-aware PCPB scheduling approach proposed in [11] to distribute the workload in the system.

3. On the basis of the daily workload statistics, to determine a set of scaling patterns describing the state of the computer system (in particular the number of active nodes of the system) such that:

    - The probability of job loss is minimized and is less than what is required by SLA.

    - Workload is processed with sufficient performance.

    - Total energy consumption of the system is minimized. Individual functions of power consumption are used to define the total energy consumption of the system.

Scale the system horizontally with the respect of defined patterns. Patterns should be formed only once on the basis of long-term workload statistics.

4. To detect the deviation of the current workload from the statistical one and adjust scaling patterns dynamically for the fulfillment of the conditions listed above.

4.1 Individual $P_j = f(CPU_j)$ dependencies definition

As the first step of the approach, it is proposed to define individual power consumption functions $P_j = f(CPU_j)$ for each node of the system. This process is described in details in [12]. In a nutshell, it is done as follows:

- An appropriate stress test is chosen in order to load each node's CPU from 0 to 100%.

- A power consumption of each node is measured for a set of CPU utilization levels (this can be done using special hardware (multimeter/wattmeter) or software that also evaluates power consumption according to computer hardware models. It is recommended to use hardware measurement tools as they provide higher measurement accuracy).

- An analytical function is obtained from the experimental data using interpolation [15].

As a result, $P_j = f(CPU_j)$ functions are obtained in a form of polynomials (Eq. (1)).

$$P_j(CPU_j) = a * CPU_j^4 + b * CPU_j^3 + c * CPU_j^2 + d * CPU_j^1 + e * CPU_j^0 \quad (1)$$

where a, b, c, d and e are fourth degree polynomial coefficients, defined within interpolation process. An example of the $P_j = f(CPU_j)$ curves obtained for five computing nodes is presented in **Figure 4**.

Having $P_j = f(CPU_j)$ functions and performance values $flops_j$ (in FLOPS) for all nodes in the system, we can range them from the worst one to the best one as follows:

1. For each node calculate the area under power consumption curve using Eq. (2).

$$S_j = \int_0^{100} P_j(CPU_j)dCPU_j, \tag{2}$$

In Eq. (2) CPU load is expressed as a percentage from 0 to 100%.

2. Sort all nodes by their S_j value in descending order.

3. Sort all nodes by their $flops_j$ value (performance metric) in ascending order.

4. Grant each node with a mark that equals the sum of the node positions in two sorted arrays. The node with the highest mark is considered to be the best one.

Thus, using this simple approach, we are able to form sorted list of nodes in order to make a decision, in which the node is the most inefficient one and should be switched off first during scale-in process (**Figure 4**). Obviously, this only makes sense for inhomogeneous systems where nodes are different from each other and can be sorted.

4.2 Basic scheduling scheme

As a basic scheduler in a system, it is proposed to use PCPB energy-aware scheduler that was proposed and further developed in [11–13]. The main idea of this

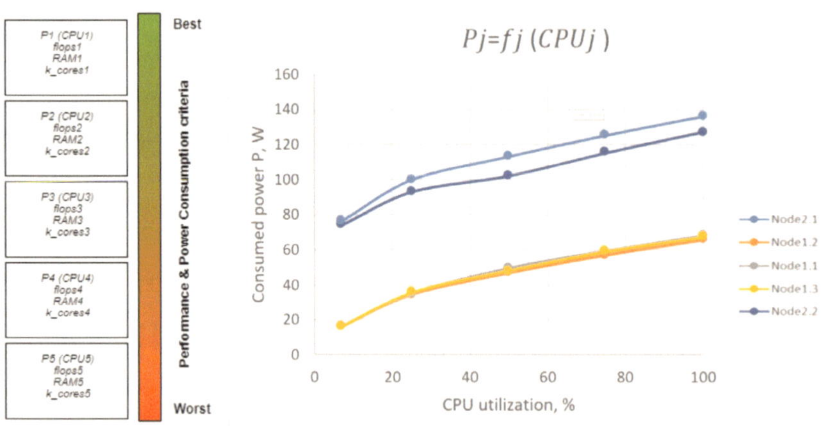

Figure 4.
Sorted list of five nodes according to their integrated performance and power consumption criteria and their $P_j = f(CPU_j)$ *curves obtained experimentally.*

scheduler is to use individual power consumption functions $P_j = f(CPU_j)$ within the scheduling process in order to determine dynamically, on which node the task is going to consume less power. This approach considers performance criteria as well. The main idea of approach is similar to that one described for the nodes' ranging; however, in PCPB the current state of computing nodes is considered. As it possible to see in **Figure 4**, the increase in energy consumption at different levels of CPU load is different (the curves grow steeper when the load changes from 0 to 25% and flatter when the load changes from 75 to 100%).

The PCPB scheduling algorithm is described by **Figure 5**.

Scheduler gathers the data regarding current load of the nodes in a system. In Step 1 it evaluates the CPU utilization of each node at the time moment τ_{k-1}, before scheduling the next job in a queue.

Having the knowledge on how many resources the job_i requires, scheduler excludes those nodes that currently do not have enough RAM or cores available for the execution (Step 2).

In Step 3 scheduler calculates the theoretical total power that is going to be consumed by the whole computing system supposing that job_i is given for the processing to the node N_j in the moment τ_k (Eq. (3)):

$$P_{\Sigma_i}|\tau_k \& N_j = P_\Sigma|\tau_{k-1} + \Delta P_{ij}|\tau_k, \tag{3}$$

where $P_\Sigma|\tau_{k-1}$ is the total power consumption of the whole system at the moment τ_{k-1} and $\Delta P_{ij}|\tau_k = f(CPU_j|\tau_k)$ is the increase of power consumption, in the case of job_i being allocated to the node N_j at the moment τ_k.

After that the nodes are sorted according to the current $P_{\Sigma_i}|\tau_k \& N_j$ values and according to their performance value $flops_j$. The nodes get the mark for their position in both sorted lists. The job is allocated to the node with the maximal mark (the detailed description of PCPB approach is presented in [11]).

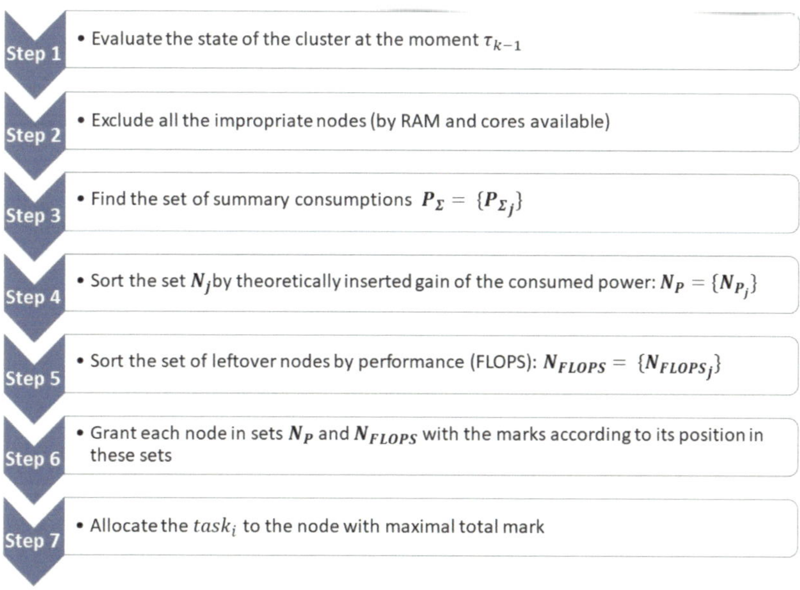

Figure 5.
PCPB scheduling algorithm.

4.3 System state pattern determination

Since within previous research [13], it was determined that energy-aware scheduling gives the best results in terms of energy efficiency while being used in conjunction with horizontal scaling; let us define the optimal system scale patterns in a form of set $\{n, t_{start}, t_{end}\}$, where n is a number of active computing nodes in the system and $\{t_{start}, t_{end}\}$ denotes the boundaries of time interval of the day, during which these patterns remain optimal. These patterns are formed with respect to the power consumption and job loss probability metrics. In order to find an optimal pattern for each time interval, it is a need to determine these two metrics as functions of number of active nodes in the system.

The certain level of system availability for service user is guaranteed by the SLA as the probability that service will be available (i.e. the user's job will be served) when it is requested. Let the probability of the job loss required by SLA be defined as in Eq. (4):

$$p_{loss_{SLA}} = 1 - p_{SLA} \tag{4}$$

where P_{SLA} is guaranteed by SLA probability that the job will be served.

The more nodes available in the system, the less likely it is that a new incoming task will find the system in a state where all nodes and the queue are fully loaded. It means that the probability of job loss depends on the number of available nodes in computing system and can be presented as a function $p_{loss} = f(n)$, where n is the number of active nodes in the system.

Thus, the purpose of scaling patterns is to answer the question: How many active nodes does the system need to have in order to fulfill the requirements to p_{loss} and minimize total power consumption P_{Σ}?

In order to estimate the dependency between the probability of job loss p_{loss} and number of active nodes n, consider the queueing system with a finite queue. The basic concepts of queuing theory are requests (or customers) and servers [16]. The natural way of modeling considered system as a queueing system would be to treat jobs as requests and computing nodes as servers. However, in this case we would have to eliminate the fact that the job may be served by several cores of one server and one server may serve several jobs simultaneously as well. These facts bring the certain complexity to the queueing system modeling. To be more precise, consider one computing core to be a server in terms of queueing system. In order to model input workload as requests of queueing system, we introduce a correction factor $k = cores_{\min avg}$ that denotes how many cores in average are requested by an input job. Using this correction factor, we can represent an input workload as a number of queueing system requests per time unit [Eq. (5)]:

$$\lambda(t) = k * \lambda_{stat}(t), \tag{5}$$

where $\lambda_{stat}(t)$ is the statistical workload represented as a number of computational jobs.

Input data for queuing system:

- Number of active servers in the system:
- n computing nodes or $n' = \sum_{j=1}^{n} k_{core_j}$ servers in terms of queueing system, where k_{core_j} is the number of cores of jth node.

- Queue length: Q jobs or $k * Q$ queueing system requests

- Service discipline:

Serving discipline is basically defined by scheduler that is being used. In considered case, using PCPB, the jobs are selected from the queue according to the first-in-first-out (FIFO) discipline.

- *The mean arrival rate:*

The mean arrival rate can be defined on the basis of the workload statistics as it is done in Eq. (5).

However, in order to build a model, it is worthy to operate with the value λ_{Tmax} that is the maximal value of load during the period T (T defines the time of one pattern validity).

- *The mean service rate:*

In general case, in order to evaluate the service rate for a particular server of the system, it is necessary to determine the average number of requests that leave the server after a successful processing per time unit. This is a random value; within this study we will consider service rate to be determined experimentally (on the basis of statistic of load processing for a particular system) and defined by its mean value μ_{avg}. In this case, the model is actually simplified, but still allows us to estimate the general relationship between the probability of loss and the number of nodes in the system.

Queueing system problem:

To define the probability of job loss, $p_{loss} - f(n)$ as the function of the number of active servers n (nodes of computing system), such that $2 < n < N$, where N is the total number of nodes in the system.

This system can be represented with a Markov chain (**Figure 6**), where S_0 denotes the state of the system, when 0 servers are occupied and the queue is empty, $S_1 - S_{n'}$ denotes there are 1 *to* n'occupied servers, and $S_{n'+1} - S_{n'+kQ}$ denotes all servers are occupied and some requests are placed to a queue.

The probability of loss for the queueing system with a finite queue is defined by Eq. (6) [16]:

$$p_{loss} = p_0 * \frac{\rho^{n'+kQ}}{n'!n'^{kQ}} \tag{6}$$

where

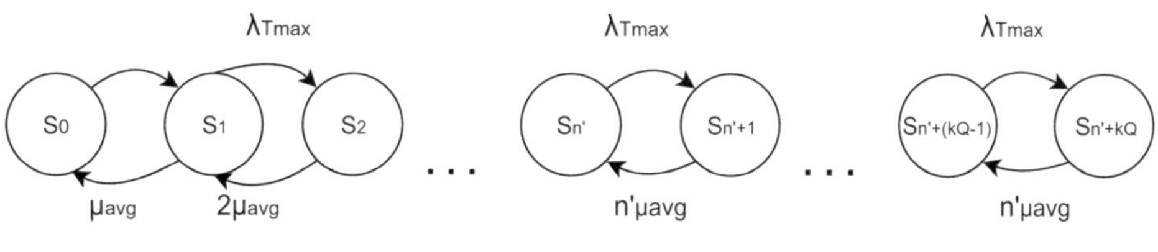

Figure 6.
System model as a Markov chain.

$$p_0 = \left(1 + \frac{\rho}{1!} + \frac{\rho^2}{2!} + \dots + \frac{\rho^{n'}}{n'!} + \frac{\rho^{n'+1}}{n' * n'!} * \frac{\left(1 - \left(\frac{\rho}{n'}\right)^{kQ}\right)}{1 - \frac{\rho}{n'}}\right)^{-1}$$

$$\rho = \lambda_{Tmax} / \mu_{avg}$$

n'—number of servers in queueing system.

kQ—length of the queue.

Thus, the simplified model described above allows us to estimate the dependency between the number of nodes of the system and probability of the loss. This model should be precisely adjusted to each real system, but the general principle may still stay the same.

Let us estimate the power consumption of the system in order to find an optimal number of nodes so that the power consumption is minimized and probability of loss does not exceed the permissible by SLA value.

As the first step, we defined individual functions of power consumption $P_j = f(CPU_j)$ for each node in the system. These functions may be formally represented in the form of polynomials (Eq. (1)).

The total power consumption of the whole system (as a sum of nodes' individual power consumptions) depends on the load that is being processed by the nodes. We have the daily workload on the system as an input data. However, according to the system model, current load of each system node highly depends on the scheduling technique.

Initially, input workload is represented as a "number of jobs per time unit". Each job may differ computationally and can be evaluated by its maximal execution time or job volume (as a number of floating-point operations). For the sake of simplicity, let us consider the static situation in some point of time, when we have k jobs that would totally fit to the processing nodes. Assume that these jobs in total form the load to the system of $\sum_{i=1}^{k} V_i [operations]$. In this case, scheduler will schedule these jobs accordingly to its policy. In the simplest case, if round-robin fair scheduler is used [17], it would schedule tasks in a way that each node would get an average $s = \frac{\sum_{i=1}^{k} V_i}{n}$ of load. Let s be a scheduler coefficient, which should be defined for each concrete scheduler. In the case of using round-robin, the dependency of power consumption of the node from its load can be formulated as Eq. (7):

$$\begin{aligned} P_j = f(CPU) = f\left(\frac{\sum_{i=1}^{k} V_i}{n} * C\right) = a * \left(\frac{\sum_{i=1}^{k} V_i}{n} * C\right)^4 + b * \left(\frac{\sum_{i=1}^{k} V_i}{n} * C\right)^3 + c * \\ * \left(\frac{\sum_{i=1}^{k} V_i}{n} * C\right)^2 + d * \left(\frac{\sum_{i=1}^{k} V_i}{n} * C\right)^1 + e * \left(\frac{\sum_{i=1}^{k} V_i}{n} * C\right)^0, \end{aligned} \tag{7}$$

where C is a constant, which expresses how the input load is converted into a CPU load in percentages. And for the more general case, Eq. (7) is transformed into Eq. (8):

$$\begin{aligned} P_j = f(CPU) = f(s * C) = a * (s * C)^4 + b * (s * C)^3 + \\ + c * (s * C)^2 + d * (s * C)^1 + e * (s * C)^0 \end{aligned} \tag{8}$$

This formula is very simplified though. Depending on scheduler being used, the load of the system's nodes may be different.

There is another more generalized approach to evaluate power consumption of system nodes. Usually, servers are kept to be utilized at some average level CPU_{avg_j} that may be defined statistically for each system. It is actually more precise solution in comparison with analytical definition of CPU utilization that depends on scheduling being used and incoming load. For the whole system, total power consumption is then defined by Eq. (9):

$$P_\Sigma = \sum_{j=1}^{n} P_j\left(CPU_{avg_j}\right), \tag{9}$$

where $P_j\left(CPU_{avg_j}\right)$ is the individual functions of nodes' power consumption.

The mathematical models of total power consumption and loss probability mentioned above lead us to the optimization problem that can be formulated as follows:

To define an optimal number of active nodes of the distributed computing system n, the objective function (Eq. (10)) is minimized subject to Eq. (11).

$$P_\Sigma = \sum_{j=1}^{n} P_j\left(CPU_{avg_j}\right) \rightarrow min \tag{10}$$

subject to:

$$p_{loss} = p_0 * \frac{\left(\frac{\lambda_{Tmax}}{\mu_{avg}}\right)^{n'+kQ}}{n'!n'^{kQ}} < p_{loss\,SLA}, 1 < n < N \tag{11}$$

To solve this problem, we need to choose the time intervals T for which we will create the patterns. Since patterns are created statically on the basis of statistical workload, it is possible to define the time moments, when thresholds are achieved so that $p_{loss} = p_{loss\,SLA}$. When it happens there is a need to increase the number of nodes (thus, to change the pattern). In case the load is decreasing, it is possible to define the moment when P_Σ is not minimal anymore for the current statistical load.

The above problem solved for each time interval T would allow to determine the optimal number of processing nodes n for each time interval depending on the incoming load. Meanwhile, the following conditions will be fulfilled:

1. Energy consumption of the system will be minimal.

2. Job loss probability will meet the SLA requirements.

On the basis of the obtained values of n, T, it is possible to create patterns for the number of nodes of the system for periods of the day. These patterns may be created once for the regular workload and be corrected during further system operation. Since the dependencies of energy consumption and loss probability from the number of active nodes in system are already defined, it requires only minimal effort to adapt the system to a new workload pattern.

4.4 Deviations from patterns detection and system reconfiguration

In order to cope with unpredictable service workload deviations that can be caused by different reasons (e.g. holidays, special events, etc.), there is a need to detect significant deviations from the statistic workload curve, predict the workload for the next chosen time period and reconfigure the system, respectively.

The authors of the research [18] have proposed an approach to hybrid resource provisioning in virtualized networks. The main idea of this approach is as follows:

- To use workload statistics to create a baseline resource allocation scheme
- To monitor deviation of the actual workload from the statistical one
- To react dynamically to deviations from the base workload, which exceeds the permissible value
- To predict the workload for the next time interval T and adjust the resource allocation accordingly

This approach can be briefly explained using **Figure** 7, where an example of applying the approach is highlighted by red circle.

Within the current study, this approach may be adapted to serve the need of detecting deviations from the statistical workload in considered distributed systems and to adjust the number of active system nodes in accordance with defined optimization problem.

Consider that the system has k scaling patterns that correspond to the daily workload (**Figure 8**). Pattern in this context means the number of active nodes that remains optimal for the certain time period, so the pattern may be denoted as a set $\{n, t_{start}, t_{end}\}$, where n is the number of active nodes and t_{start}, t_{end} show the interval of the day, during which these patterns remain optimal. These patterns may contain other configuration information as well, but it lies beyond the scope of the current study.

According to the adapted approach proposed in [18], the current workload should be monitored throughout the day, but the intensity of the monitoring may vary according to the deviation between the statistical workload and the current one. Let I_{base} be a baseline time monitoring interval. This interval should be determined empirically for each system. Monitoring interval then should be adjusted according to Eq. (12):

$$W(t) = I_{base} - K \sum_{j=t-h}^{t-1} \frac{\max(0; \lambda_{obs}(t) - \lambda_{pred}(t))}{h} I_{base}, \quad (12)$$

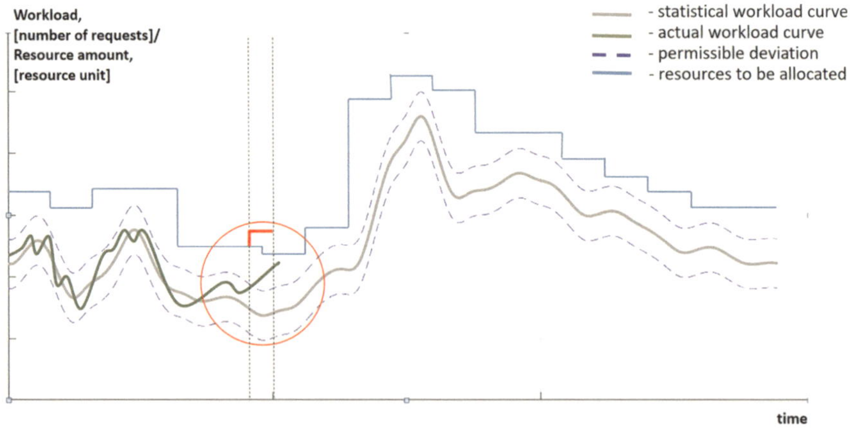

Figure 7.
The main idea of hybrid resource provisioning approach [18].

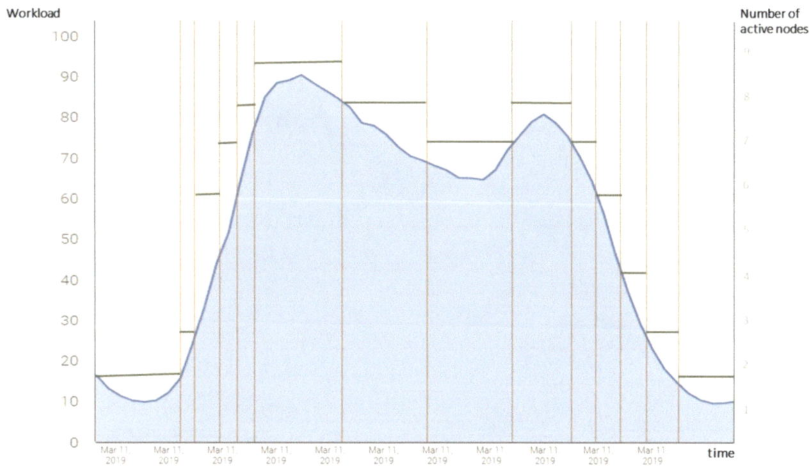

Figure 8.
Scaling patterns of the system that correspond to the daily workload.

where:

W is an new monitoring interval.

I_{base} is the baseline predefined monitoring interval.

K is the normalization constant that should be defined for each system individually.

$\lambda_{obs}(t)$ is the current load arrival rate during the time t.

$\lambda_{pred}(t)$ is the predicted load arrival rate during the time t (according to the statistical workload curve).

h is the number of preceding intervals considered by the algorithm.

If $\lambda_{basepred}(t)$ is a basic (statistical) load arrival rate for a period of time t and $\lambda_{obs}(t)$ is the current load arrival rate during the time t, the load predicted for the next interval should be adjusted according to Eq. (13):

$$\lambda_{pred}(t) = \lambda_{basepred}(t) + \sum_{j=t-h}^{t-1} \frac{\lambda_{obs}(j) \quad \lambda_{pred}(j)}{h} \tag{13}$$

This approach is described in details in [18].

Thus, having the predicted value of workload for the next time period, we can recalculate the pattern for that period and change the number of nodes in the system according to the calculated optimal number.

As a result of applying proposed steps, we get formed patterns $\{n, t_{start}, t_{end}\}$ for the static day workload and may adjust them according to the deviation between current workload and statistical one.

5. Modeling

Within the modeling the system with comparatively low workload and number of computing nodes was chosen for clarity, but such model can be scaled for larger systems as well. Matlab system was used to create the model.

Suppose we have a daily workload curve depicted in **Figure 9** (the value of workload is defined for each minute, so we also have value λ for each minute).

For this workload let us define static scaling patterns that describe the optimal numbers of active nodes and time periods, for which these numbers remain optimal.

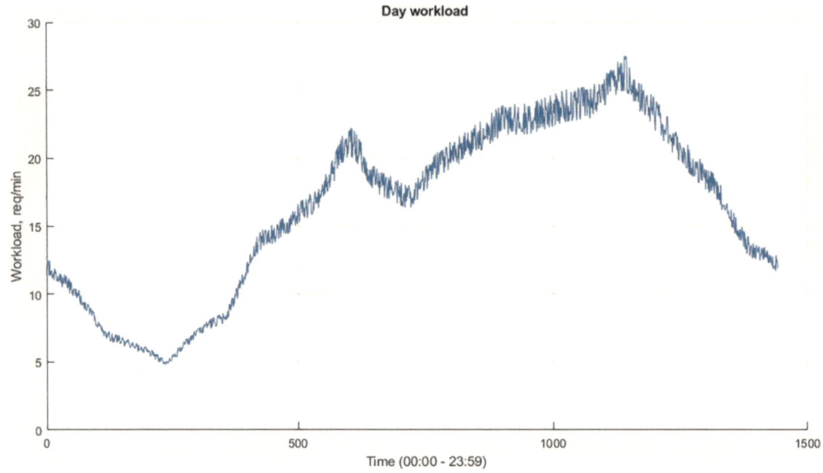

Figure 9.
Daily workload curve used for modeling.

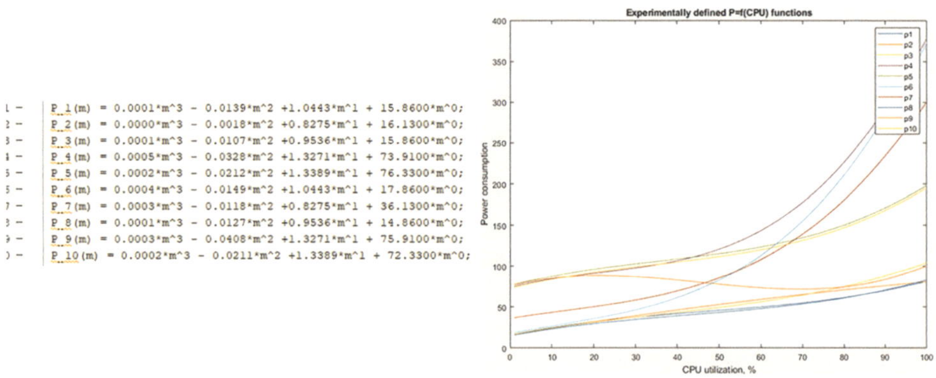

Figure 10.
Analytical and graphical representation of the $P = f(CPU)$ functions defined experimentally.

As an input data for modeling, we also have the following:

1. $P = f(CPU)$ functions for $N = 10$ servers in the form of polynomials. These functions were defined experimentally and interpolated using fourth degree polynomials. Analytical and graphical representation of the $P = f(CPU)$ functions is presented in **Figure 10**.

2. Corresponding values of RAM volume, number of cores and performance for these ten servers.

3. Scheduler algorithm that is being used—PCPB algorithm (the details regarding PCPB algorithm modeling are provided within the previous research [11]).

4. Average serving rate for given servers: $\mu_{avg} = 25$ req/min.

5. $Q = 5$, the length of the queue.

6. Let the probability of loss be defined by SLA be $p_{lossSLA} = 10^{-6}$.

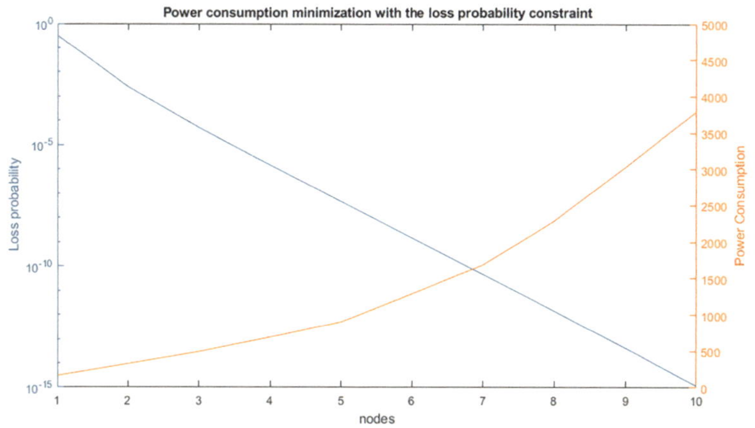

Figure 11.
Loss probability and power consumption calculations for the first workload value.

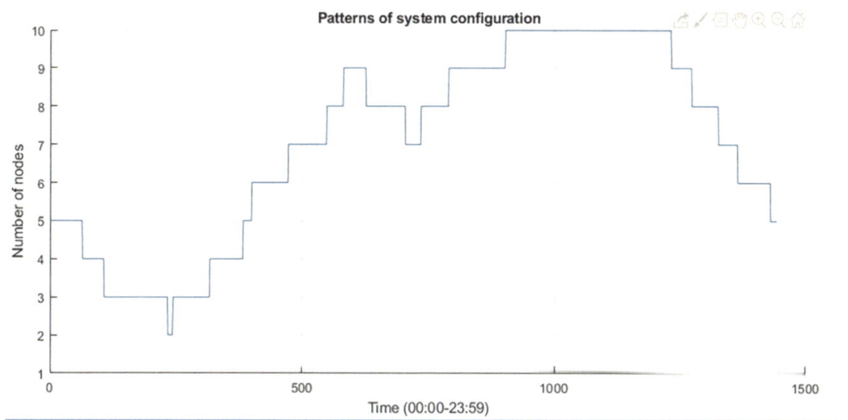

Figure 12.
Created scaling patterns for the given input workload.

In order to create patterns for certain time periods, it is worthy to run calculations according to Eq. (1) for each minute at first. Then, as soon as the thresholds of p_{loss} and $P_{j\Sigma}$ are defined, it is possible to derive $\{t_{start}, t_{end}\}$ values for each pattern.

Let us build the curves described by Eq. (10) and Eq. (11) for the first minute workload. Resulting curves are depicted in **Figure 11**.

In **Figure 11** it is possible to see that in order to meet the SLA requirements $p_{loss SLA} = 10^{-6}$, we need to have at least five nodes in the system. We can also see that these five nodes will consume near 900 W of energy. As the nodes are sorted from the worst one to the best one, these five leftover nodes are the most energy efficient and productive among the others.

We need to do the same for the remaining part of workload curve. This process was modeled using Matlab. As a result, we got all the scaling patterns for the given input workload (**Figure 12**).

Let us introduce some deviation to the statistical workload in order to apply pattern adjustment. Introduced workload deviation is circled in red in **Figure 13**. Adjustment was made according to Eqs. 12 and 13 proposed in [18].

In **Figure 13**, the adjustment of the received patterns is shown in the right graph. As it is possible to see, the interval of patterns optimality and number of nodes were changed, respectively.

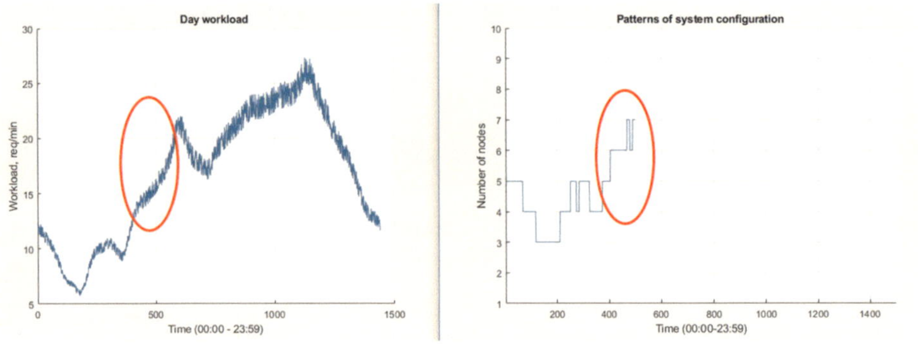

Figure 13.
Scaling patterns' dynamic adjustment result.

Thus, created model fully illustrates the process of proposed intelligent workload scheduling approach which is improved through the use of scaling approaches. The model shows the main idea of proposed approach; however, in real systems it is worthy to add some redundancy and always keep at least one extra server active in order to cope with unpredicted workload deviations.

6. Conclusion

In this chapter an intelligent approach to the workload scheduling in distributed computing environment was proposed. According to the proposed approach, energy-aware PCPB scheduling algorithm is combined with scaling approaches that allow to achieve an optimal balance between energy efficiency and performance while fulfilling the SLA requirements. In order to achieve these goals, we propose to create and dynamically adjust system scaling patterns while using energy-aware scheduling. An approach was modeled using Matlab. The simulation results showed that it is able to cope with the efficient processing of statistical load, as well as load deviations.

Within the future research, the system representation as a queueing system is going to be made more precisely, and proposed approach's efficiency should be proven by means of experiment.

Author details

Larysa Globa[1*], Oleksandr Stryzhak[2], Nataliia Gvozdetska[1] and Volodymyr Prokopets[1]

1 National Technical University of Ukraine "Igor Sikorsky Kyiv Polytechnic Institute", Kyiv, Ukraine

2 National Academy of Sciences in Ukraine, Kyiv, Ukraine

*Address all correspondence to: lgloba@its.kpi.ua

References

[1] Gartner. Gartner Top 10 Strategic Technology Trends for 2019 [Internet]. 2018. Available from: https://www.gartner.com/smarterwithgartner/gartner-top-10-strategic-technology-trends-for-2019/

[2] Andrae A, Edler T. On global electricity usage of communication technology: Trends to 2030. Challenges. 2015;**6**:117-157. DOI: 10.3390/challe6010117

[3] Beloglazov A. Energy-efficient management of virtual machines in data centers for cloud computing [thesis]. Department of Computing and Information Systems, The University of Melbourne; 2013

[4] Suleiman D, Ibrahim M, Hamarash I. Dynamic voltage frequency scaling (DVFS) for microprocessors power and energy reduction. In: Proceedings of the 4th International Conference on Electrical and Electronics Engineering. 2005

[5] Akram S, Sartor J, Eeckhout L. DVFS performance prediction for managed multithreaded applications. In: Proceedings of the IEEE International Symposium on Performance Analysis of Systems and Software-ISPASS. 2016. pp. 12-23

[6] Agarwal K, Nowka K. Dynamic power management by combination of dual static supply voltages. In: Proceedings of the 8th International Symposium on Quality of Electronic Design (ISQED 2007); 26-28 March 2007; San Jose, CA, USA. 2007

[7] Möbius C, Dargie W, Schill A. Power consumption estimation models for processors, virtual machines, and servers. In: Proceedings of the IEEE Transactions on Parallel and Distributed Systems. 2014. pp. 1600-1614

[8] Aldossary M, Djemame K. Performance and energy-based cost prediction of virtual machines auto-scaling in clouds. In: Proceedings of the 44th Euromicro Conference on Software Engineering and Advanced Applications (SEAA 2018); 29-31 August 2018; Prague, Czech Republic. 2018. pp. 502-509. DOI: 10.1109/SEAA.2018.00086

[9] Liu W, Du W, Chen J, Wang W, Zeng G. Adaptive energy-efficient scheduling algorithm for parallel tasks on homogeneous clusters. Journal of Network and Computer Applications. 2013;**41**:101-113. DOI: 10.1016/j.jnca.2013.10.009

[10] Armenta-Cano F, Tchernykh A, Cortés-Mendoza J, Yahyapour R, Drozdov A, Bouvry P, et al. Heterogeneous job consolidation for power aware scheduling with quality of service. In: Proceedings of the Russian Supercomputing Days. RuSCDays'15; Moskow. 2015

[11] Schill A, Globa L, Stepurin O, Gvozdetska N, Prokopets V. Power consumption and performance balance (PCPB) scheduling algorithm for computer cluster. In: Proceedings of the 2017 International Conference on Information and Telecommunication Technologies and Radio Electronics (UkrMiCo 2017); Odesa, Ukraine. 2017. pp. 1-8

[12] Gvozdetska N, Stepurin O, Globa L. Experimental analysis of PCPB scheduling algorithm. In: Proceedings of the 14th International Conference the Experience of Designing and Application of CAD Systems in Microelectronics (CADSM); 21-25 February 2017; Polyana-Svalyava (Zakarpattya), Ukraine. 2017

[13] Bezruk V, Globa L, Stryzhak O, editors. Knowledge-Based Technologies

of Optimization and Management in Infocommunication Networks: Monograph. 1st ed. Kyiv: National Academy of Pedagogical Sciences of Ukraine; 2019. p. 194. ISBN 978-617-7734-02-3

[14] Google. Google Transparency Report [Internet]. 2019. Available from: http://www.google.com/transparencyreport/traffic/

[15] Mastroianni G, Milovanovic G. Interpolation Processes: Basic Theory and Applications. 1st ed. Berlin, Heidelberg: Springer; 2008. p. 446. DOI: 10.1007/978-3-15540-68349-0

[16] Sztrik J. Basic Queueing Theory. 1st ed. Debrecen, Hungary: University of Debrecen, Faculty of Informatics; 2011. p. 193

[17] Wikipedia. Round-Robin Scheduling [Internet]. 2019. Available from: https://en.wikipedia.org/wiki/Round-robin_scheduling

[18] Sulima S, Skulysh M. Hybrid resource provisioning system for virtual network functions. Radio Electronics, Computer Science, Control. 2017;(1): 16-23. DOI: 10.15588/1607-3274-2017-1-2

Chapter 4

Approximation for Scheduling on Parallel Machines with Fixed Jobs or Unavailability Periods

Liliana Grigoriu

Abstract

We survey results that address the problem of non-preemptive scheduling on parallel machines with fixed jobs or unavailability periods with the purpose of minimizing the maximum completion time. We consider both identical and uniform processors, and also address the special case of scheduling on nonsimultaneous parallel machines, which may start processing at different times. The discussed results include polynomial-time approximation algorithms that achieve the best possible worst-case approximation bound of 1.5 in the class of polynomial algorithms unless P = NP for scheduling on identical processors with at most one fixed job on each machine and on uniform machines with at most one fixed job on each machine. The presented heuristics have similarities with the LPT algorithm or the MULTIFIT algorithm and they are fast and easy to implement. For scheduling on nonsimultaneous machines, experiments suggest that they would perform well in practice. We also include references to the relevant work in this area that contains more complex algorithms. We then discuss the main methods of argument used in the approximation bound proofs for the simple heuristics, and comment upon current challenges in this area by describing aspects of related practical problems from the automotive industry.

Keywords: multiprocessor scheduling, availability constraints, fixed jobs, uniform processors, worst-case approximation, nonsimultaneous machines, makespan, maximum completion time, unavailability

1. Introduction

The necessity to assign resources such as machines to jobs that need to be performed without interruption, where the time required for a machine to perform a certain job is known in advance, is a widely encountered problem. It can occur for example in production planning or when assigning landing and take-off stripes to planes in airports. Sometimes these resources become unavailable for predetermined periods of time, for example because of necessary maintenance. Minimizing the maximum completion time of all tasks is often considered as a goal, for example such that the workers who operate the machines can undertake other activities afterward or go home early. As a consequence, the problem of scheduling on multiple machines with predefined unavailability periods (downtimes) to minimize the maximum completion time, that is, the latest completion time of a job in a schedule, has been considered. A closely related problem, of scheduling with fixed jobs, where on each machine certain jobs have to be performed at predefined times, has also been considered. The difference between these two problems is in the meaning of the objective function: when scheduling with fixed jobs, the maximum completion time of the jobs must be at least the latest completion time of a fixed job, whereas the maximum completion time when scheduling in the presence of unavailability periods can occur before the end of an unavailability period. We consider the static nonresumable variant of the problem of scheduling with unavailability periods, where the downtimes are known for each machine before the schedule needs to be made, and where jobs that start executing before a downtime cannot resume execution after it.

In these problems, the job execution times are usually assumed to be given as an integer number of computing units such as clock cycles or of other suitable units such as time units. Similarly, the starttimes and endtimes of unavailability periods or of fixed jobs are assumed to be given as integer multiples of adequately chosen time units. We note that any problem with rational numbers as job durations and starttimes and endtimes of downtimes or of fixed jobs can be transformed into an equivalent problem where these entities are integers by multiplying them with an adequate factor, thus there is arguably no loss of generality in this assumption when considering the representation of any practical problem.

Both the problem of multiprocessor scheduling on fixed jobs and that of multiprocessor scheduling with unavailability periods are strongly NP-hard as they are more general than the strongly NP-hard multiprocessor scheduling problem (MSP), which has no downtimes or fixed jobs.

For scheduling with downtimes, it is NP-hard to find a schedule that ends within a given constant multiple of an optimal schedule even when scheduling on identical machines with at most one downtime on each machine. We discuss this in more detail in Section 4.2. To obtain approximation results for scheduling with unavailability periods in this context, assumptions about the downtimes were made such as the assumption that only a fraction of the processors can be unavailable at the same time [1, 2], or by comparing the generated schedule to the latest among the end of an optimal schedule or the latest end of a downtime, thus essentially considering scheduling with fixed jobs [3, 4].

To describe the performance of an approximation algorithm, we use the notion of a *worst-case approximation bound*. In this work, we call worst-case approximation bound of an algorithm A when applied to a scheduling problem SP the largest ratio between the maximum completion time of a schedule produced by A and the maximum completion time of an optimal schedule for a problem instance of SP.

For the problem of multiprocessor scheduling with fixed jobs to minimize the maximum completion time, even in the case where there is at most one fixed job on each machine, it has been shown in [5] that no polynomial algorithm can achieve a worst-case approximation bound that is less than 1.5 unless P = NP. Sharbrodt et al. [5] also give a polynomial-time approximation scheme (PTAS) for scheduling on a constant number of uniform processors with fixed jobs. Polynomial-time approximation algorithms for this problem that achieve the worst-case approximation bound of 1.5 were given for the general problem in [6]. For the case where there is at most one fixed job on each machine, significanlty simpler heuristics with lower time complexities resembling the largest processing time algorithm (LPT) [7] for identical processors and the MULTIFIT algorithm [8] for uniform processors also achieve this bound [3, 4].

The case where all downtimes are at the beginning of the processing time on all machines is called scheduling with nonsimultaneous machine available times, as the machines start processing jobs nonsimultaneously. For this problem, polynomial-time algorithms with constant worstcase approximation bounds exist.

For scheduling on identical nonsimultaneous parallel machines, MULTIFIT achieves a tight worst-case approximation bound of 24/19 (~1.2632) [9] and another algorithm achieves a bound of 5/4 [10], while in the case of scheduling on uniform nonsimultaneous parallel machines, a MULTIFIT variant has a worst-case approximation bound of 1.382 [11], which was shown by generalizing the bound obtained for MULTIFIT when scheduling on uniform processors in [12]. Experimental results suggest that for scheduling on nonsimultaneous uniform machines, the MULTIFIT variant from [11] is adequate for use in practice, as we discuss in Section [4].

In Section 2, we describe the ways in which the content of this work can be used. In Section 3, we introduce the algorithms LPT and MULTIFIT. In Section 4, we consider scheduling with unavailability periods, while first focussing on scheduling with nonsimultaneous machine available times in Section 4.1, and on the more general case where downtimes do not have to occur at the beginning of the schedule in Section 4.2. In Section 5, we present results on scheduling with fixed jobs. Section 6 contains the description of main techniques used in the worst-case approximation bound proofs and Section 7 contains concluding remarks and a discussion of some of the challenges in this area.

2. Contributions of this work

We next present ways to use the content of this work.

2.1 A deeper understanding

This work aims to provide a deeper understanding of multiple related problems that involve scheduling on parallel machines with fixed jobs or unavailability periods to minimize the maximum completion time. We explain why multiprocessor scheduling with at most one unavailability period on each machine does not have a polynomial-time approximation algorithm with a constant worst-case approximation bound unless P = NP, which is the main reason why results on this topic are hard to obtain.

Furthermore, we observe that most results in this area involve variants of LPT and MULTIFIT, and comment on the other results obtained. We also hope that this work will increase awareness of these results and of how they relate to each other.

2.2 Practical use of the heuristics

The heuristics presented and referenced in this work can be used directly in practice or for research purposes to solve the problems they address. The heuristics based on LPT and MULTIFIT are fast and easy to implement and some of them have best possible worst-case approximation bounds in the class of polynomial algorithms unless P = NP for the problems they address. In addition to worst-case approximation results, this work also highlights for some cases experimental insights into how the heuristics would perform in practice based on how they perform for randomly generated instances. As expected, they perform much better for such instances than in the worst case. Also, for some cases, references to more complex methods are provided in case the user prefers to use those.

2.3 Proof techniques

This work presents the main proof techniques used to obtain worst-case approximation bounds for LPT- and MULTIFIT-like heuristics when the aim is to minimize the maximum completion time. Thus, the interested reader is provided with a concise description of the tools that can be used to develop such proofs, and he or she may not have to read hundreds of pages in order to become aware of all of them or work with an expert in the area when developing such a proof. Even for people with expertise in the area, one or more of the ideas presented may be new and helpful.

3. The algorithms LPT and MULTIFIT

The algorithms LPT and MULTIFIT are among the most studied approximation algorithms for multiprocessor scheduling with or without unavailability periods or fixed jobs. In this section, we describe the basic versions of these algorithms for MSP, which can be stated as follows: given a set of m machines P and n jobs J find a non-preemptive schedule that ends at the earliest possible time. A non-preemptime schedule is an assignment of jobs to processors, together with an order in which the jobs on each processor are processed.

The algorithm LPT [7] works as follows:

Algorithm 1 The largest processing time algorithm (LPT)

1: Order jobs in nonincreasing order of their processing time.
2: In this order assign each job at the earliest possible time allowed by the schedule that exists when the job is assigned.

The algorithm MULTIFIT was first introduced by Coffmann Garey and Johnson in 1978 [8]. It uses binary search for the end of its resulting schedule and receives as input an accuracy ε with which it determines this schedule end. In each iteration it assigns a deadline and attempts to create a schedule that contains all tasks that ends at or before that deadline by using the bin packing algorithm first fit decreasing (FFD). If a feasible schedule is created, it decreases the deadline and otherwise it increases the deadline. This process is repeated until the difference between the current deadline and the previously considered deadline is less than ε. More formally, the algorithm is described as Algorithm 2.

The MULTIFIT algorithm results in a schedule with a maximum completion time that is within ε of the maximum completion time of the schedule that would result if the binary search for the deadline would be continued.

An example of a LPT-schedule and a MULTIFIT schedule for the same problem instance are presented in **Figures 1** and **2** respectively.

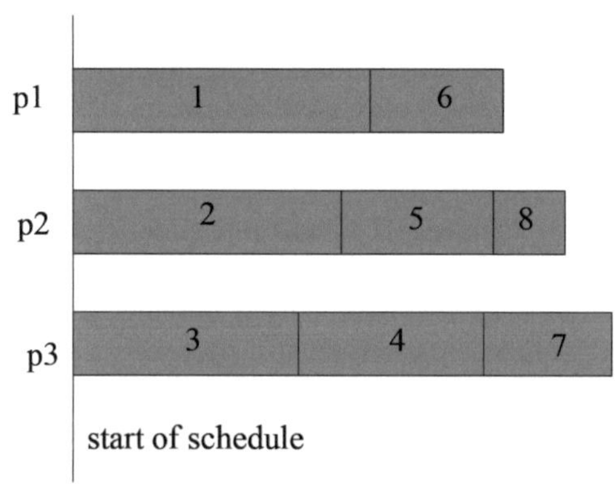

Figure 1.
A LPT schedule. The jobs are numbered according to the order in which they are considered. At start, when all processors are available at the same time, they are considered in the order p1, p2, p3 in this example.

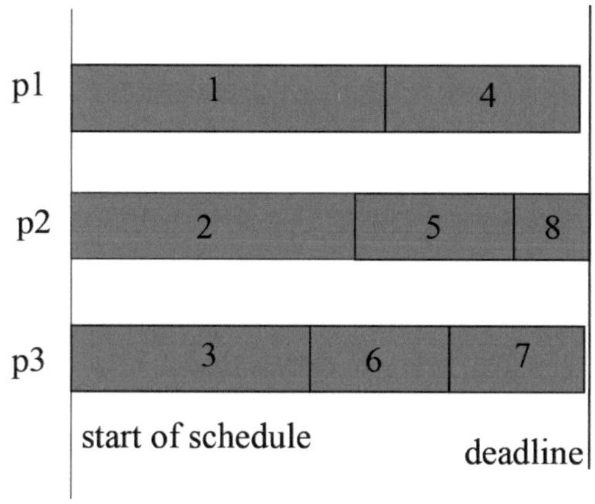

Figure 2.
A MULTIFIT schedule together with a possible deadline. The jobs are numbered according to the order in which they are considered. The processors are considered in the order p1, p2, p3 when generating the schedule.

Algorithm 2 The algorithm MULTIFIT

1: Order the jobs in non-increasing order of their duration.
2: Assign upper bound (ub) and lower bound (lb) for the end of schedule; (for example, lb = sum of job durations/number of processors, ub = sum of job durations).
3: Assign b = (ub + lb)/2 as deadline.
4: FFD: Assign tasks in non-increasing order on the first processor on which they fit while respecting the deadline (the processors are considered in each iteration in the same order).
5: If all tasks are successfully scheduled decrease the upper bound: ub = b.
6: Else increase the lower bound: lb = b.
7: If ub - lb ≥ ε loop back to Step 3.

The MULTIFIT algorithm tends to produce more balanced schedules than LPT, and, as a consequence it tends to perform better when the aim is to minimize the maximum completion time. It also has a higher time complexity, as it tries to make a schedule about $\log_2(ub - lb)$ times, whereas the LPT algorithm only does that once. When the instances are prohibitively big and the schedule needs to be made in a short time, it may be indicated to use LPT or another list scheduling algorithm to schedule the jobs. This is because in practical situations, there may not be enough time to go through the list of jobs more than once while scheduling thousands of jobs and making sure that all required constraints are obeyed by the schedule. The reason why such big schedules are made is that the companies aim to estimate delivery times for their orders.

3.1 Time complexity of MULTIFIT

If the parameter ε of MULTIFIT is adequately set, for example as a computer clock cycle, the algorithm returns the best possible MULTIFIT schedule, that is, running it further would not result in a better schedule, as was commented upon in [4].

The binary search for the deadline within the MULTIFIT algorithm happens within $\log_2[(ub - lb)/\varepsilon]$ time, which is at most $\log_2(ub/\varepsilon)$, which is the size of ub in binary, assuming that the numbers for the upper bound and the lower bound do not change their representation during the execution of the algorithm and that they allow within their representation for an accuracy of ε. In Section 1, we mentioned that the job durations are usually given as integer multiples of an adequately chosen (time) unit; therefore, the end of any schedule is an integer, and thus, there is no point in making ε less than 1, in which case the MULTIFIT loop is repeated $\log_2(ub)$ times.

As a consequence, the number of times the MULTIFIT loop is called is polynomial in the size of the input, as any reasonable upper bound is at most the sum of the processing times of all jobs, which can be represented within at most the total number of bits used to represent all jobs. In [4], Grigoriu and Friesen also comment that if the upper bound is 2 years, the lower bound is 0 and the deadline is determined with an accuracy of 10^{-13} s, the MULTIFIT loop is called at most 40 times.

The time complexity of MULTIFIT is thus $O(n \log n + nm \log_2(ub))$, as the jobs need to be sorted according to their execution times in a non-increasing order in Step (1), and as in each iteration of the MULTIFIT loop, the algorithm looks for each job for the first processor on which it fits; thus, it will have to look at most at m processors. Recall that n is the number of jobs in the considered problem instance.

4. Scheduling with unavailability periods

In this section, we first present results for the case where all unavailability periods are at the beginning of the schedule. Then, we present results for the more general case where the unavailability periods can occur anywhere in the schedule.

4.1 Scheduling with nonsimultaneous machine available times

This section addresses the case where the processors may have unavailability periods at the start of their processing time. This situation is more general than the multiprocessor scheduling problem (where there are no fixed jobs or downtimes) and less general than the problems of scheduling with fixed jobs or with unavailability periods. As the less general multiprocessor scheduling problem is NP-hard, so are the problems of scheduling on identical machines with nonsimultaneous machine available times (NMSP: nonsimultaneous multiprocessor scheduling problem) and scheduling on uniform processors with nonsimultaneous machine available times (UNMSP) when minimizing the maximum completion time. Due to the NP-hardness of these problems, polynomial-time approximation algorithms like LPT and MULTIFIT and their variants have been studied for their solution. As before, we will continue to denote with the number of processors in the problem instance being considered with m.

4.1.1 Scheduling on identical nonsimultaneous processors

For NMSP, worst-case approximation bounds for LPT of $3/2 - 1/(2m)$ and for a modified version of LPT (MLPT) of 4/3 have been obtained by Lee [13]. The bound obtained by Lee in [13] was improved upon by Kellerer in [10], where a dual approximation algorithm with a worst-case approximation bound of 5/4 was presented.

When MULTIFIT is used for MSP, a deadline results in periods of equal duration in which jobs can be scheduled on each processor; thus the schedules resulting from using any ordering of processors in step (4) of MULTIFIT have the same maximum completion time. When considering NMSP, thus allowing for nonsimultaneous machines, the order in which processors are considered becomes relevant, as the period in which jobs can be executed on each processor corresponding to a deadline depends on the time the processor becomes available. MULTIFIT variants that address such situations usually order the processors in non-decreasing order of their periods in which jobs can be scheduled. Thus, in this case, the ordering is in non-increasing order of the times at which the processors become available.

A bound of 9/7 (about 1.2857) was obtained for MULTIFIT by Chang and Hwang [14]. In [10], Kellerer gives a problem instance of NMSP for which the approximation factor of its MULTIFIT schedule is 24/19 (about 1.2632). More recently, 24/19 was shown to be the exact worst-case approximation bound when using MULTIFIT for NMSP by Hwang and Lim [9]. By comparison, a tight worst-case approximation bound of 13/11 (about 1.18182) was shown by Yue [15] for MUTLIFIT when applied to MSP.

4.1.2 Scheduling on uniform nonsimultaneous processors

For the uniform multiprocessor scheduling problem with simultaneous machine available times (UMSP), that is, where processors execute jobs at different speeds, the amount of jobs that fit on a processor corresponding to a given MULTIFIT deadline depends on the speed of that processor. Usually, the slowest processor is considered to have a speed of 1, and for each job j, the time it would take to process it on this processor l_j is given. Thus, on a machine with speed 5, a job j needs a time of $l_j/5$ to be processed. As a consequence, the ordering in which the processors are considered in Step (4) of MULTIFIT is in most cases relevant to the maximum completion time of the resulting schedule. MULTIFIT for UMSP orders processors in each iteration before its Step (4) in non-decreasing order of the duration of the processing time on that processor times the speed of that processor [12, 16].

For UMSP, approximation bounds of 1.4 and 1.382 were obtained for MULTIFIT by Friesen and Langston [16] and by Chen [12] respectively. In [17], Burkard and He derive a worst-case approximation bound of $\sqrt{6}/2$ (about 1.2247) of MULTIFIT for UMSP with at most two processors, and show a better bound of $(\sqrt{2}+1)/2$ (about 1.2071) for the case where MULTIFIT is combined with LPT as an incumbent algorithm.

In [11], Grigoriu and Friesen show that bounds that apply to the MULTIFIT variants from previous work such as [12, 16, 17] where scheduling on two uniform processors is considered also apply to a slightly different proposed variant of MULTIFIT for UNMSP, LMULTIFIT, which was first proposed in [4] in a more general form. The difference between the MULTIFIT variants from [12, 16, 17] on the one hand and LMULTIFIT on the other hand is that in the latter, the choice of the initial upper and lower bounds is not given explicitly within the algorithm, and thus the worst-case approximation bound proofs are more general, as they work for any initial choices of upper and lower bounds for the duration of the resulting schedule. A first step in the proofs that the bounds hold in the more general case, where there are uniform nonsimultaneous parallel machines, was to show that LMULTIFIT obeys the bounds of the earlier MULTIFIT variants in the simultaneous machines case for the instances considered in those works.

Using LPT for UNMSP has been considered in [18], where worst-case approximation bound of 5/3 was shown in the general case, as well as a better bound for the case where there are only two machines.

For the case where the number of machines is constant, a PTAS exists for UNMSP [11], which was derived from a PTAS for scheduling on a constant number of uniform processors with fixed jobs from [5]. As the objective function for scheduling with fixed jobs that are at the beginning of the schedule and scheduling with unavailability periods that are at the beginning of the schedule differ, the PTAS from [5] can not be used unaltered to address UNMSP. To address UNMSP, the PTAS from [5] is first run for the transformed problem instance where the unavailability periods become fixed jobs, and then for all problem instances resulting from successively removing the machine with the latest end of a fixed job from the transformed instance [11]. This accounts for the cases where a number between 1 and $m - 1$ of processors start processing after the end of the optimal schedule.

In [19], a lower bound is derived for the end of an optimal schedule of an UNMSP instance, and using that bound approximation factors for LMULTIFIT schedules of randomly generated instances are determined. The reasonably extensive experiments described in [19] suggest that LMULTIFIT is a good option for solving UNMSP in practice, not only because of being fast and easy to implement, but also because it has very good approximation factors (less than 1.03) for the generated instances with an average of at least five jobs for each machine. In order to obtain the approximation factors, a lower bound for the end of the optimal schedule that can be calculated directly from the problem instance was used.

4.2 Multiprocessor scheduling with availability constraints

In this section, we consider the multiprocessor scheduling problem where downtimes can occur at any time during the scheduling horizon.

Surveys with focus on scheduling with availability constraints are given in [20–24]. Besides the makespan, the authors of these works survey work on various other objective functions such as total completion time, and also address additional variants of the problem, such as its resumable version.

Unless assumptions about the unavailability periods are made or unless P = NP, there is no polynomial-time approximation algorithm with a constant worst-case approximation bound for the problem of scheduling with unavailability periods to minimize the maximum completion time, since there is a polynomial-time reduction from the NP-hard problem of 3-Partition to the problem of finding a schedule that has a maximum completion time that is at most α times the end of an optimal schedule for this problem. We next outline such a reduction. Let X be an instance of 3-Partition, that is, a set of $3m$ positive integers, given with the purpose of finding out whether there is a partition of these numbers into m sets, such that the sum of the numbers in each set is the same for all sets. Let S be the sum of all numbers in X. The instance Y of scheduling with unavailability periods that we build is given as follows: there are m processors, each of which has an unavailability period of duration $(\alpha + 1)S$ that starts at time S/m, and the job durations in Y are the numbers in X. X is a yes-instance of 3-Partition if and only if in instance Y, the optimal schedule ends at time S/m. In such a situation, any approximation algorithm with worst-case approximation factor α for scheduling with availability constraints will find a schedule that ends at or before time $\alpha S/m$ which is less than $(\alpha + 1)S$. Thus, the found schedule must end before or when the unavailability periods start, at time S/m. In such a schedule, the sets of durations of jobs on each processor are a 3-Partition of X. Therefore, any polynomial-time approximation algorithm for scheduling with unavailability periods with a worst-case approximation factor α can be used to solve 3-Partition in polynomial time.

4.2.1 Scheduling on identical machines with unavailability periods

For resumable scheduling, where the execution of jobs may continue after a downtime that interrupted them, but where jobs cannot be preempted by the scheduling algorithm, and where one machine does not shut down and all other machines shut down at most once, Lee shows that the makespan of LPT is in the worst case $(m + 1)/2$ times the optimal makespan [25].

In [1], Hwang and Chang make the assumption that at most half of the machines are unavailable at any time, and show for this situation that the worst-case approximation bound of LPT is 2. In [3], it is shown that no polynomial algorithm can have a better bound than 2 for this problem unless $P = NP$. The result from [1] is generalized in [2] where it is shown that if at most $\lambda \in \{1, \ldots, m - 1\}$ machines may be unavailable at any time, LPT has a worst-case approximation bound of $1 + \frac{1}{2}\lceil m/(m - \lambda) \rceil$, and that this bound is tight for LPT.

4.2.2 Scheduling on uniform machines with unavailability periods

In [26], scheduling with at most one unavailability period on each machine is considered and exact algorithms are given for small problem instances. The authors consider separately the case of identical jobs, and also consider total completion time beside the makespan as an objective function. For larger instances, they propose an LPT-like algorithm, which assigns jobs in nonincreasing order of their processing time to the fastest machine on which they would finish being processed at the earliest time. They also do experiments on a total number of 68 generated instances where error margins of at most 5.6% are observed. They do not compare their heuristic to the previously proposed heuristic from [4], which we discuss in Section 5.1.2, which was also proposed for this problem, even though its worst-case approximation bound was shown for the objective function of scheduling with fixed jobs.

5. Scheduling with fixed jobs

The problem of scheduling with fixed jobs is given as a number of processors, where each processor may have jobs that must be executed during certain given periods on it, together with a number of other jobs which can be executed by any processor. As noted in Section 1, job durations or required execution times are expressed for example as a number of significant units such as clock cycles. For uniform processors this number represents the time needed by a job to be executed on the slowest processor. In case there are uniform processors, each processor also has a speed factor, by which the time needed by a job on the slowest processor is divided in order to obtain the time needed for the processor to execute the job.

As noted before, the problem of scheduling with fixed jobs differs from the problem of scheduling with unavailability periods in that its maximum completion time cannot be earlier than the latest completion time of a fixed job.

In [5], Scharbrodt et al. give a polynomial-time approximation scheme for scheduling on a constant number of uniform machines with fixed jobs. They also show that it is NP-hard to obtain a schedule that ends within a factor of less than 1.5 when scheduling on identical processors with at most one fixed job on each machine. Even though they do not specify their result in this way, their proof that no polynomial-time approximation algorithm can have a better worst-case approximation bound than 1.5 for multiprocessor scheduling with fixed jobs does not use the fact that there can be more than one fixed job on each machine, which implies the previous statement.

If all fixed jobs are at the beginning of the schedule, the results presented in Section 4.1 apply, as the optimal schedule of each problem instance of scheduling on nonsimultaneous machines can only potentially get worse when scheduling with fixed jobs is considered instead, and since the resulting schedule of an approximation algorithm ends later for scheduling with fixed jobs only if its maximum completion time is the completion time of a fixed job, which the optimal schedule also needs to execute.

We next consider the case where there is at most one fixed job on each machine in Section 5.1, and the case where there can be multiple fixed jobs on each machine in Section 5.2.

5.1 Scheduling with at most one fixed job on each machine

When scheduling on multiple processors with at most one fixed job on each machine, LPT and MULTIFIT variants have been shown to achieve a worst-case approximation bound of 1.5, which is best possible for this problem unless $P = NP$.

5.1.1 Same-speed processors

For scheduling on identical machines with at most one fixed job on each machine, an LPT-like algorithm, LPTX, was given in [3], for which a worst-case approximation bound of 1.5 was shown. Before running LPT, LPTX creates an order of processors, which is then used by LPT to break ties in case two processors become available at the same time. The ordered list of processors created before applying LPT is built in two steps:

1. All processors that have an unavailability period that is not at the beginning of the schedule are assigned to this list in non-decreasing order of the start of their downtime.

2. All other processors (that is, those which have the downtimes at the beginning of the scheduling period or have no downtime at all) are appended to the list built in the previous step in non-decreasing order of the times at which they can start executing jobs.

5.1.2 Uniform processors

In the more general case of scheduling on uniform machines with at most one fixed job on each machine, a MULTIFIT-like algorithm, LMULTIFIT, was given in [4], which achieves the worst-case approximation bound of 1.5. For a MULTIFIT variant to work in the presence of downtimes, it must be specified how it deals with the fact that there are more than one time interval in which jobs can be scheduled on one processor.

After a MULTIFIT deadline is assigned and before applying the bin packing algorithm FFD (see Section 3), LMULTIFIT orders all time intervals in which jobs can be scheduled in non-decreasing order of their *length*. Here, the length of a time interval is the duration of the time interval multiplied by the speed factor of the processor on which the interval occurs.

5.2 Scheduling with multiple fixed jobs on each machine

For scheduling on identical processors with fixed jobs, where the number of fixed jobs on a machine can be arbitrary, approximation algorithms that are much more complex than LPT and MULTIFIT were given in [27], with a worst-case approximation bound of $1.5 + \varepsilon$, where ε is the parameter of a fully polynomial time approximation scheme (FPTAS) for the multiple subset sum problem, and in [6] with a worst-case approximation bound of 1.5, where a FPTAS for the multiple knapsack problem is used as a subroutine.

In [28], a very long proof is outlined that LMULTIFIT achieves a worst-case approximation bound of 1.5 when scheduling on identical processors with at most two fixed jobs on each machine. In [29], an algorithm using two MULTIFIT-like algorithms is shown to have a worst-case approximation bound of 1.625, which likely can be improved to 1.6 without excessive effort.

The time complexity of the MULTIFIT-like and LPT-like algorithms is significanlty lower than that of the algorithms from [6, 27], and they are also significantly easier to implement; however, there is little hope in our opinion that bounds better than 1.55 can be shown in the general case for such algorithms with proofs of reasonable length. Given such problems, the user must decide what is best suited for his or her needs.

6. Proof techniques

Worst-case approximation results in this research area about LPT and MULTIFIT variants mainly use proofs by contradiction in which some proof techniques appear very often. We next describe two techniques that we consider to be the most relevant, and then comment upon some other methods that are used relatively often. In the following, we call job lengths the durations of the jobs on the slowest processor.

6.1 Minimal counterexample

A very well-known proof method is that of assuming that there exists a minimal counterexample to a theorem T, that is, a problem instance for which the algorithm's schedule does not obey that theorem, with a minimal number of processors, of jobs, of downtimes (or of fixed jobs) that do not start at the beginning of the schedule, and/or of other quantities that can be minimized which are chosen to fit the statements to prove. A minimal counterexample exists whenever there is a counterexample, and thus, showing that it does not exist proves T. In order to define minimality among counterexamples, the author of the proof first chooses a partial order among the problem instances. An instance which is minimal according to this partial order among the instances for which a theorem T does not apply is called a minimal counterexample.

This method can be very powerful, because after assuming that a minimal counterexample does not obey a theorem T, many useful properties of a minimal counterexample can be derived from the fact that no *lesser* counterexample exists, which can ultimately lead to a contradiction.

Here, a lesser counterexample is a counterexample with less processors, or less tasks, or less downtimes, or with a job with a smaller length, depending on how the order of instances was defined. Showing that a minimal counterexample does not exist is usually significantly easier than developing a direct proof for T.

The theorem to prove could be that the worst-case approximation bound holds, but it can also be an intermediary result that is later used to prove the worst-case approximation bound. One could address instances that have a certain property first, and then show that the worst-case approximation bound holds for these, for example by using a minimal counterexample among these instances, and then do the same for all other instances.

Sometimes it is enough to define the partial order only using the number of processors [11], while in most cases, it is useful to include multiple characteristics of problem instances, such as all or a part of the characteristics enumerated above. In one situation, a minimal counterexample was defined to also have minimal job lengths, meaning that if in a minimal counterexample the length of one job is reduced, the resulting instance is not a counterexample [28].

6.2 Weighing arguments

For a problem instance that does not obey a worst-case approximation bound, there is a job (X) that is scheduled such that it crosses the bound to prove times the end of an optimal schedule (B) for LPT-like algorithms, or that can not be scheduled when the deadline is at B for MULTIFIT-like algorithms. If the order defined on instances includes the number of jobs, there is only one such job in a minimal counterexample to (for example) a theorem that the analyzed algorithm (A) generates only schedules that obey the bound. Otherwise, all jobs that would be scheduled afterward can be removed and a lesser counterexample could be obtained, resulting in a contradiction. As a consequence of how LPT and MULTIFIT work, X is the smallest job of the minimal counterexample in both cases.

The schedule S_A generated by an LPT-like algorithm A until a job would cross B and the schedule S_A generated by a MULTIFIT-like algorithm A if the MULTIFIT deadline is at B do not contain the job X. An optimal schedule, however, contains all jobs, including X.

Therefore, the optimal schedule has more total execution time than S_A. Also, if nonzero weights are assigned to each job, the optimal schedule has a total weight of all its jobs that is greater than that of S_A, the difference being the weight of X. An adequate assignment of weights to jobs can lead to the conclusion that the sum of weights of all jobs contained in the schedule S_A is at least the sum of the weights of all jobs in the considered optimal schedule, a contradiction.

There are infinite ways of assigning weights, and there is no unique strategy that leads to success. Usually, the weight function is monotonic with regard to job lengths, and, as X has the smallest job length, its weight can be set to 1. In the following, we denote both the job X and the time X would need to be executed on the slowest processor by X. The other weights can be assigned for intervals of job lengths, for example, a job with length within $[X,1.5X)$ could be assigned weight 1. Weights can also be assigned otherwise: for example, a job with a certain property can be chosen to be the end of a weight interval. Jobs that have a certain property can also be assigned a specific weight that corresponds to that property. Of course, while proving different theorems that lead to the proof of a worst-case approximation bound, a new weight function can be chosen for each statement to prove.

6.3 Normalizing time intervals and job execution times

In order to reason easier about time, one can divide all durations of time intervals in which jobs can be scheduled by X if scheduling on identical processors. For uniform processors, such intervals can be divided by the time it would take X to execute on the processor on which the interval occurs, in order to derive the *length* of the interval [4]. Also, all job durations can be divided by X and these normalized durations can be used in the proofs. For example, a theorem could be proved that there are no jobs that have a longer duration than 5, or it can be stated that the unused time intervals on processors before the MULTIFIT deadline B all have length less than 1, as otherwise X would have been scheduled in one of those intervals.

6.4 Task density

In the case of uniform processors, the time intervals can have unbounded lengths because the speed factors may be arbitrarily high. A way to describe the amount of jobs assigned within a schedule in such an interval is to use *task densities*, which can be defined for each task as being the ratio between its weight and its length. Also, a *task type density* can be defined as a lower bound for all possible task densities of tasks of that type. The concept of task density can be used in order to reason about time intervals that may be very long. For example, the total weight of all tasks in an interval of length t is at most t times the maximum task type density among all task types represented within that interval.

6.5 Processor with more execution time in the optimal schedule

We use the notations from the previous subsection. Since X is not contained in S_A, there must be a processor p^* in the optimal schedule that has a total execution time that is greater than that of S_A on p^*. Such a processor can be analyzed in detail and may be shown to have certain properties that, in conjunction with other methods, result in proofs of useful theorems. For example, the existence of p^* may imply a certain minimum duration of the optimal schedule, in conjunction with the observation that X could not be fitted by A on p^* without causing the maximum completion time of S_A to exceed B.

7. Conclusions and future challenges

In this work, we considered worst-case approximation for scheduling on multiple machines with availability constraints or with fixed jobs in order to minimize the maximum completion time. We surveyed results obtained in this research area and commented upon the algorithms used.

Prominent among these algorithms are LPT and MULTIFIT and their variants, whereas for multiprocessor scheduling with fixed jobs, more complicated algorithms were used to achieve best possible worst-case approximation bounds in the class of polynomial algorithms assuming that $P \neq NP$ in the general case where there can be any number of fixed jobs on each machine.

The problem of scheduling with availability constraints cannot be approximated by a polynomial-time algorithm with a constant worst-case approximation bound, even if there is at most one downtime on each machine, unless assumptions about the downtimes are made. The results we presented in this area address the problem of scheduling on identical processors with at most one downtime on each machine, with various assumptions.

Due to its different objective function, the problem of scheduling on identical parallel machines with fixed jobs allowed for the development of a polynomial-time approximation algorithm with a worst-case approximation bound of 1.5, and the development of a PTAS for scheduling on a constant number of uniform machines with fixed jobs was also possible.

The MULTIFIT and LPT variants developed for the discussed variants of these problems could be useful in practice, as their time complexity is low and thus they should be able to address very large problem instances, as they are easy to implement, and because in some cases their worst-case approximation bounds could be considered to be good enough. In the case of scheduling on uniform nonsimultaneous machines, the average performance of a MULTIFIT variant was studied, and shown to be very good, as the experiments suggest that in general, for instances that can be relevant in practice and for which exhaustive search is not an option, the algorithm returns schedules with a maximum completion time that is within 3% of that of an optimal schedule.

We also elaborated on the most encountered proof techniques in worst-case approximation bound proofs for LPT and MULTIFIT variants.

The limitations of the presented works result mainly from the difficulty of the problem of scheduling with unavailability periods when considering the subject of approximation. To assess how well the proposed heuristics for this problem perform under such conditions is difficult, as it is hard to have a good estimate of the optimal schedule unless it is computed by an exact algorithm as was done in [26]. This problem has therefore been addressed only by considering the special case where there is at most one downtime on each machine.

In the future, it may be interesting to compare the heuristics proposed for the same problems experimentally.

Another limitation of the discussed works and of many other research works on scheduling results from the fact they attempt to understand problems with one, two, or at most three aspects at one time, whereas in many practical problems such as some production planning problems, many aspects occur at once. For example, availability constraints can appear alongside a multitude of other constraints that have to be considered simultaneously. These can be precedence constraints, that certain jobs have to be assigned to certain machines, or preferences of the manufacturer that the machines should not have more than a 60% or another predefined load for example in order to leave room for unexpected events. Furthermore, orders often come online, and if an urgent order from an important client needs to be given priority, this can alter the delivery times of other orders. Also, delivery times and delays have a big relevance in practice, as not delivering on time can cause fines. Such practical problems can also have sequence-dependent setup times, the necessity for setup operators to be present to perform setups, the preference that setup times are kept low by putting jobs from the same family of types of jobs consecutively on machines whenever possible, the necessity for workers to attend to certain machines while production takes place, and worker breaks and holidays. The preexisting schedule also has to be kept unchanged for a predefined time period since materials are brought to the production place in preparation for the production process. In addition, orders may have priorities and deadlines. For such problems, given the time constraints in which the schedule needs to be generated and that there can be thousands of jobs, usually a heuristic is employed that first orders the jobs for example by using priorities assigned to them and/or their deadlines and then schedules them on the machines in that order while also obeying all constraints and attempting to fulfill all preferences. Difficulties in researching such problems include that probably for different sets of orders different scheduling strategies may be better, and that an optimal schedule may be very hard to find and thus it is hard to quantify how well a heuristic performs.

Acknowledgements

This publication was supported by the Open Access Publication Fund of Technische Universität Berlin.

Conflict of interest

The author declares no conflict of interest.

Abbreviations

MSP	multiprocessor scheduling problem
NMSP	nonsimultaneous multiprocessor scheduling problem
UNMSP	uniform nonsimultaneous multiprocessor scheduling problem
LPT	largest processing time
FFD	first fit decreasing

Author details

Liliana Grigoriu[1,2]

1 Department of Computer Science and Engineering, Faculty of Control and Computers, Politehnica University Bucharest, Bucharest, Romania

2 Department of Mathematics, Technical University of Berlin, Berlin, Germany

*Address all correspondence to: liliana.grigoriu@cs.pub.ro

References

[1] Hwang H-C, Chang SY. Parallel machines scheduling with machine shutdowns. Computers and Mathematics with Applications. June 1998;**36**:21-31

[2] Hwang H-C, Lee K, Chang SY. The effect of machine availability on the worst-case performance of LPT. Discrete Applied Mathematics. April 2005;**148**:49-61

[3] Grigoriu L, Friesen DK. Scheduling on same-speed processors with at most one downtime on each machine. Discrete Optimization. November 2010;**7**:212-221

[4] Grigoriu L, Friesen DK. Scheduling on uniform processors with at most one downtime on each machine. Discrete Optimization. November 2015;**17**:14-24

[5] Scharbrodt M, Steger A, Weisser H. Approximability of scheduling with fixed jobs. Journal of Scheduling. November 1999;**2**(6):267-284

[6] Jansen K, Pradel L, Schwarz UM, Svensson O. Faster approximation algorithms for scheduling with fixed jobs. In: 17th Conference of Computing: The Australasian Theory Symposium (CATS 2011), Perth, Australia, January. 2011

[7] Graham RL. Bounds on multiprocessing timing anomalies. SIAM Journal of Applied Mathematics. March 1969;**17**:416-429

[8] Coffman EG Jr, Garey MR, Johnson DS. An application of bin-packing to multiprocessor scheduling. SIAM Journal on Computing. February 1978;**7**:1-17

[9] Hwang HC, Lim K. Exact performance of MULTIFIT for nonsimultaneous machines. Discrete Applied Mathematics. 2014;**167**:172-187

[10] Kellerer H. Algorithms for multiprocessor scheduling with machine release times. IIE Transactions. 1998;**30**:991-999

[11] Grigoriu L, Friesen DK. Approximation for scheduling on uniform nonsimultaneous parallel machines. Journal of Scheduling. December 2017;**20**:593-600

[12] Chen B. Tighter bound for multifit scheduling on uniform processors. Discrete Applied Mathematics. May 1991;**31**:227-260

[13] Lee CY. Parallel machine scheduling with nonsimultaneous machine available time. Discrete Applied Mathematics. January 1991;**30**:53-61

[14] Chang SY, Hwang HC. The worst-case analysis of the MULTIFIT algorithm for scheduling nonsimultaneous parallel machines. Discrete Applied Mathematics. June 1999;**92**:135-147

[15] Yue M. On the exact upper bound of the MULTIFIT processor scheduling algorithm. Annals of Operations Research. December 1990;**24**:233-259

[16] Friesen DK, Langston MA. Bounds for multifit scheduling on uniform processors. SIAM Journal on Computing. February 1983;**12**:60-69

[17] Burkard RE, He Y. A note on MULTIFIT scheduling for uniform machines. Computing. 1998;**61**:277-283

[18] He Y. Uniform machine scheduling with machine available *constraints*. Acta Matematicae Applicatae Sinica (English Series). 2000;**16**:122-129

[19] Grigoriu L, Friesen DK. Scheduling on uniform nonsimultaneous parallel machines. In: Fink A, Fiigenschuh A, Geiger M, editors. Operations Research

Proceedings 2016 Selected Papers of the Annual International Conference of the German Operations Research Society (GOR), Hannover, August 30–September 2. 2016. pp. 467-473

[20] Lee CY, Lei L, Pinedo M. Current trends in deterministic scheduling. Annals of Operations Research. April 1997;**70**:1-41

[21] Sanlaville E, Schmidt G. Machine scheduling with availability constraints. Acta Informatica. September 1998;**35**:795-811

[22] Schmidt G. Scheduling with limited machine availability. European Journal of Operational Research. February 2000;**121**:1-15

[23] Lee C-Y. Machine scheduling with availability constraints. In: Leung JY-T, editor. Handbook of Scheduling: Algorithms, Models and Performance Analysis. London: Chapman & Hall/CRC; 2004. pp. 22-1-22-13

[24] Ma Y, Chu C, Zuo C. A survey of scheduling with deterministic machine availability constraints. Computers & Industrial Engineering. 2010;**58**:199-211

[25] Lee CY. Machine scheduling with an availability constraint. Journal of Global Optimization. December 1996;**9**:395-416

[26] Kaabi J, Harrath Y. Scheduling on uniform parallel machines with periodic unavailability constraints. International Journal of Production Research. 2019;**57**(1):216-227

[27] Diedrich F, Jansen K. Improved approximation algorithms for scheduling with fixed jobs. Proceedings of 20th ACM-SIAM Symposium on Discrete Algorithms (SODA). 2009:675-684

[28] Grigoriu L. Multiprocessor scheduling with availability constraints [PhD thesis]. College Station, TX, USA: Texas A&M University; 2010

[29] Grigoriu L. Scheduling on parallel machines with variable availability patterns [PhD thesis]. Bucharest, Romania: Politehnica University Bucharest; 2012

Chapter 5

An Empirical Survey on Load Balancing: A Nature-Inspired Approach

Surya Teja Marella and Thummuru Gunasekhar

Abstract

Since the dawn of humanity man tried to mimic several animals and their behavior be it in the age of hunting of while designing the aero plane. Human brain holds a significant amount of power in observing the species around him and trying to incorporate their behavior in several walks of life. This mimicking has helped human to evolve into beings which we are now. Some typical examples include navigation systems, designing several gadgets like aero planes, boats, etc. These days these inspirations are several, and their inspiration is being utilized in several fields like operations, supply-chain management, machine learning and several other fields. The similar kind of approach has been discussed in this paper where we tried to analyze different phenomenon in nature and how different algorithms were designed from these and how these can ultimately be used to solve different issues in cloud balancing. Essential component of cloud computing is load balancer which holds a crucial role of task allocation in virtual machines and several kinds of algorithms were developed on different ways of task allocation procedures each holding its significance here we tried to find the optimal resource allocation in terms of task allocation and rather than approaching through traditional methods we tried to solve this issue by using soft computing techniques. Specifically, nature-inspired algorithms as it hold the key to unlocking massive potential regarding research and problem-solving approach. The central idea of this paper is to connect different optimization techniques to load balancer and how could we make a hybrid algorithm to serve the purpose. We also discussed several different types of algorithms each bearing its roots from different natural procedures. All the algorithms in this paper can be broadly tabulated into three different types SO (Swarm optimization techniques), GO (Genetic-based algorithms), PO (Physics-based algorithms).

Keywords: load balancing, cloud computing, nature-inspired algorithms, optimization techniques

1. Introduction

Load balancing implies guaranteeing the even distribution of workloads and to adjust the load among the accessible resources ideally. It helps in accomplishing a high client fulfillment and asset utilization proportion. Many scheduling algorithms have been proposed to maintain load balancing. The primary point is the ideal assets utilization resulting in improved throughput, migration times or smaller response, ideal adaptability, and overall system production [1]. There are individual difficulties in cloud computing, and that needed to be routed to give the most reasonable and productive effective load balancing algorithms. These challenges are:

1. Geographical/spatial distributions of the nodes: to design an LBA that works for spatially or geographically distributed nodes is an arduous task. It is because as the distance increases the speed of the network links among the nodes is influenced which thusly influences the throughput [2].

2. The complexity of algorithm: complexity affects the overall performance of a system. Generally, LBA has a less complicated implementation. Complexity lead to delays which further causes more problems [3].

3. Point of failure: the load balancing algorithm ought to be planned in a way that they abstain from having a single point of failure.

4. Static load balancing algorithm: a static load balancing algorithm works on the earlier state/previous data, not on the ongoing state. It cannot adjust to the load changes at run-time [4].

The challenges as mentioned above remain unsolved. Researchers are going on to bring about changes in the existing algorithms and to create new algorithms to overcome these challenges.

Cloud load balancing is the activity of regulating the workload and computing assets in a cloud computing environment to accomplish high performance at potentially lower costs. This includes facilitating the dissemination of workload activity and requests that dwell over the Interweb [5]. As we know that, a load balancing technique is basically appropriating workloads among the servers and processing assets in a cloud domain in which the number of clients where more significant than the servers so that there can be the burden on the servers so we need to balance the load so we distribute the tasks among the servers equally so it cannot be the bash with any other server and in this way we can increase the performance of a server [6]. By allocating the resources among the various computer network or server, Load balancing allows companies or organizations to manage the applications or workload demands, so load balancing in cloud computing that incorporate facilitating the distribution of workload activity and request over the system so first level every one of the customers have been composed, In second level all the servers have been organized and in between these two a load balance rare used to balance the load among the server and is generally used by the company or organizations to manage their applications. Cloud computing is an advanced worldview to give benefits through the Interweb [7]. Load balancing is an essential part of cloud computing and avoids the situation in which a couple of nodes wind up finished weight while the others are sitting still or have little work to do. Load balancing can upgrade the Quality of Service (QoS) estimations, including response time, cost, throughput, execution and asset utilization. In computing, Load balancing [8] enhances the distribution of workloads over various figuring assets, for example, PCs, a PC gathering, central processing units, network links, or disk drives. Load balancing means to enhance asset usage, maximize throughput, confine response time, and maintain a strategic distance from over-weight of any single asset. Using different parts with load balancing rather than a single component may increase reliability and accessibility through excess. The load balancing in clouds may be among physical hosts or VMs. This balancing segment scatters the dynamic workload fairly among every one of the hubs (hosts or VMs).

The heap adjusting in the cloud is additionally alluded to as load adjusting as an administration (LBaaS) [9] (**Figure 1**).

Load adjusting is to move the workload to computational assets that are underutilized, with an outrageous objective of lessening the general execution time. A considerable measure of research has been added to the point, and this case continues with framework figuring and disseminated computation [10]. In the area of multi-core computing, a typical multi-core framework comprises of same cores that communicate using shared memory space. Similarly stays consistent with GPUs also [11]. Thus, a notwithstanding dividing of the load among accessible cores should do the trick to deliver a base execution time. Regardless, this dispute is effectively countered by the manner in which that we can have a gathering of workloads, each with various or obscure computation requirements. An even or approach apportioning would not be doable. If a distributed memory system is utilized as an execution stage, correspondence overheads can turn into a genuine performance concern [11].

As a rule, legitimately dividing the workload is basic to boosting execution. Toward this objective, we ought to think about a comparable number of the stage (e.g., computational speed) and issue attributes (e.g., cost of data transmission) as conceivable [12]. Dynamic load adjusting insinuates a wide assembling of computations that perform or change stack assignments on the Web, i.e., in the midst of

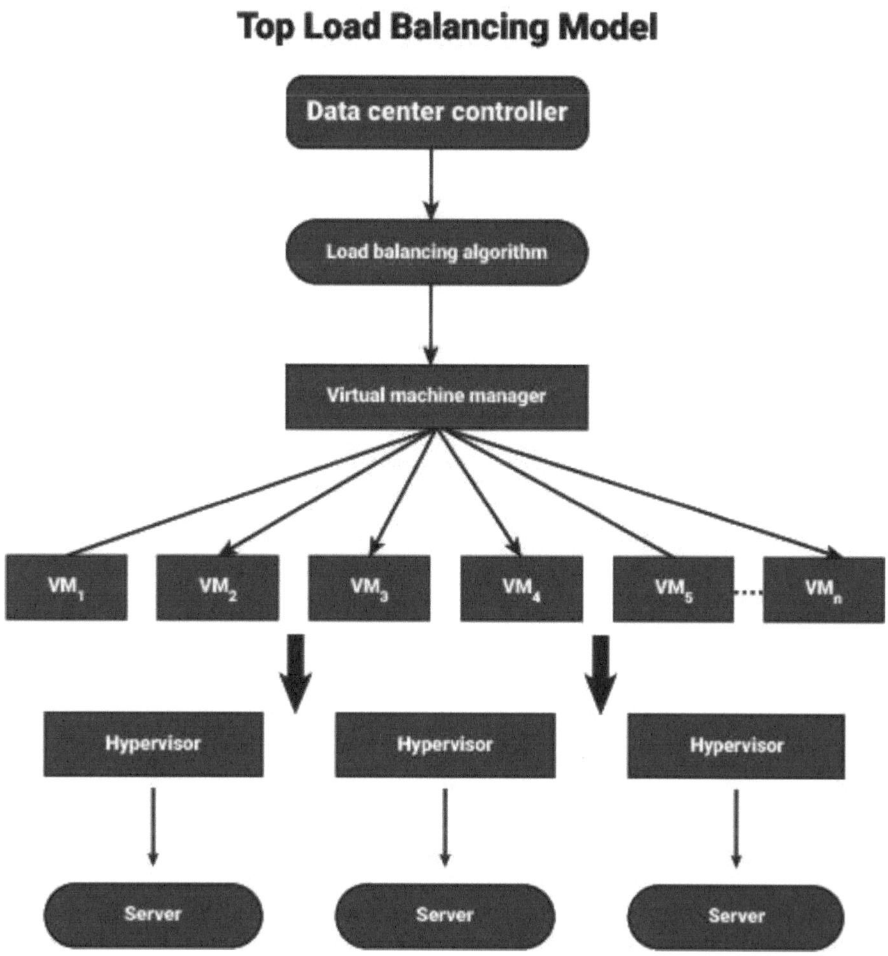

Figure 1.
Load balancing model.

the execution of a program [13]. Dynamic load balancing is depicted by the ability to adjust to changes to the execution organize (e.g., hubs going isolates, correspondence joins finding the opportunity to be congested, and whatnot.) yet to the burden of additional coordination overhead [14]. Static load modifying can give a close ideal answer for the load-partitioning issue, the exchange offs being the failure to adapt to run-time changes and the need to develop insinuate learning about the execution characteristics of the individual fragments making up the execution arrange [15].

To attain these goals, two types of load balancing algorithms are designed:

1.1 Static load balancing

In static load balancing, cloud requires knowledge of processing power, performance, nodes capacity and memory [5]. The cloud additionally requires learning of client necessities which cannot be changed on run-time. It is simpler to mimic the static condition, yet in the event that client necessities change static condition cannot adjust to it. Two well-known algorithms applied in static environments are [16].

1.1.1 Round Robin algorithm

In the Round Robin algorithm, assets are allocated to errand on First-come-First-Serve (FCFS) premise. It implies the errand which arrives first; assets are distributed to it first. In this algorithm, tasks are scheduled in time sharing manner [17].

1.1.2 Central load balancing decision model

It is an improved approach to Round Robin algorithm. It uses the basics of the Round Robin algorithm. This algorithm calculates total execution time spent by a task on the cloud. It uses this information to calculate time elapsed during client and server communication [18].

1.2 Dynamic load balancing

In a dynamic environment, various resources are installed. In a dynamic environment, cloud considers runtime statistics [19]. In a dynamic environment, the cloud allows changes in user requirements on runtime. Algorithms in a dynamic environment can quickly adapt to runtime changes. The dynamic environment is challenging to simulate [20]. Various load balancing algorithms implemented in a dynamic environment are weighted least connection (WLC) algorithm, load balancing min-min (LBMM) algorithm and opportunistic load balancing (OLB) algorithm [21].

Considering the scalability and free nature of the cloud, dynamic environments are preferred over static environments for cloud implementation as it also satisfies the particular goals of load balancing as given below,

Goals of load balancing are:

- To maintain the fault tolerance of the system.
- To maintain the stability of the system.
- To improve efficiency and performance of the system.
- To minimize job execution time.

- To minimize time spent waiting in the queue.
- To facilitate improved resource utilization ratio

1.3 Related works

Each technique holds significance and a different approach to solve the problem. Swarm optimization is generally used to find an optimal solution it might not be the best fit solution, Genetic algorithms generally try to find the best fit but it consumes much time, and physics algorithm generally used as hybrid or support algorithm to minimize other procedures in different algorithms. Swarm optimization is generally inspired by observing different flock behaviors be it in frogs, bats, fireflies and ants and each takes a specific approach to the problem of food gathering, communication, foraging, etc. [22]. A genetic algorithm is approximately enlivened by Charles Darwin theory of survival of fittest and algorithms like a genetic algorithm, mimetic algorithm come under this category. Physics algorithms are inspired from different physical phenomena like gravitation, mass-energy equivalence, simulated annealing, etc. These generally act as support factors for different algorithms. In every algorithm the process can be generalized as randomization, calculating fitness and arriving at a possible solution. The approach varies, but the process is more or less the same. The underlying algorithms used these days in different spheres are ant colony optimization; in the chapter, genetic algorithm and simulated annealing are also discussed. We also discussed concepts like task allocation based on a few traditional algorithms. Factors like green computing which affect the performance of the device considerably are also discussed. To conclude we tried to solve the optimal resource allocation in a natural way because nature itself looks for best fitting procedures for its procedures and mimicking it in computing could be advantageous [25].

To expand the general execution of the framework, load balancing is an intense instrument that aides in conveying bigger workloads into smaller processing workloads. To achieve proper resource utilization and excellent user satisfaction, it helps in the fair allocation of computing resources. It avoids bottlenecks and implements failover thus increasing the scalability. To transfer and receive data without any delay, load balancing divides the traffic between all the servers to get an optimum solution. The central vision of load balancing is to make sure that at any point in time, the processors in the system does the same amount of work. It is necessary for load balancing to utilize full parallel and distributed system's resources. Load balancing is classified as dynamic load balancing and static load balancing [23]. The processor's performance is decided at the beginning of execution in static load balancing. From that point onward, as per their execution, the workload is partitioned by the ace processor. It should be possible utilizing algorithms named "central manager algorithm," "threshold algorithm," and "round robin algorithm." The work is divided during the runtime in dynamic load balancing. With new information collected by the master, new processes it assigned to the slaves. Here, processes are allocated dynamically. It can be carried out using algorithms such as, "local queue algorithm" and "central queue algorithm" [24].

A load balancer can perform the following functions:

1. Distributes network load and requests of clients across many servers.
2. According to demand, it can add and subtract the servers.
3. Provides high scalability, reliability and availability to the online servers.

To scale with the increased demands, vendors of the cloud are more toward automatic load balancing services that allow the entities to increase the memory and count of CPU for their resources.

2. Models of nature-inspired algorithms

Nature-inspired algorithm is a capacity that aroused by activities that are perceived by nature. These registering approaches prompted the change of development named nature-inspired algorithms (NIA) [25]. This breakthrough is apt for computational agility. The objective of developing such algorithms is to optimize engineering problems [26] (**Figure 2**).

These algorithms use recombination and change overseers to streamline the perplexing issues, e.g., genetic algorithm and differential evolution et cetera. The basic objective of nature-inspired algorithms is to discover a universally response for a given issue. Two key factors normal in all nature-inspired algorithms are strengthening and expansion regularly named as Exploration and Exploitation [27].

Some of the algorithms are,

2.1 Hill climbing

It is a nature-inspired algorithm which takes its inspiration from a process of hill climbing. It is an iterative algorithm. This algorithm is thus helpful to find the minimal solution and can be used in the load balancer [28]. A tool called load balancer which is used to find the assets allocation in the cluster and several kinds of algorithms are used to find the resource allocation to the cluster and here to find the solution we can use the hill climbing algorithm.

Initially, the algorithm considers the cluster as a graph and mimics hill climbing behavior on that graph and generates a solution for optimal resource allocation [29].

To locate the ideal answer for the given issue this strategy can be utilized, this method can be utilized as a part of load adjusting to locate the ideal asset assignment for the given issue.

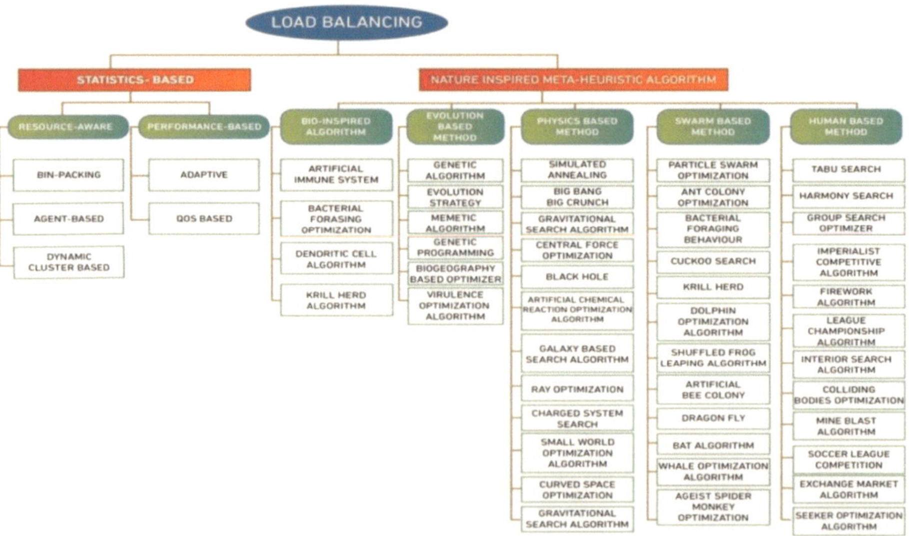

Figure 2.
Load balancing taxonomy.

This technique has three different variants,

1. **Coordinate descent**: this technique is optimally used to determine the optimal solution for the given problem but cannot be implemented as it has an exponential time worst case scenario.

2. **Stochastic hill climbing**: this technique does not guarantee an optimal solution, but it is better than stochastic hill climbing regarding the time taken to find the optimal solution.

3. **Traditional hill climbing approach**: this technique cannot be used in the load balancer as the solution we get out of this is not optimal.

Hence, here we describe a stochastic hill climbing approach as it is best regarding both time complexity and optimization (**Figure 3**).

```
Present node = initialNode;
 loop
   k=neighbours(presentNode);
nextEvalatuation = -knf
nextNode = NULL;
    for all x in L
       if (EVALuation(x) >nextEvalution)
nextNode = j;
nextEvaluation = evaLuation(x);
    if nextEvaluation<= evaluation(presentNode)

       return presentNode;
presentNode = nextNode;
```

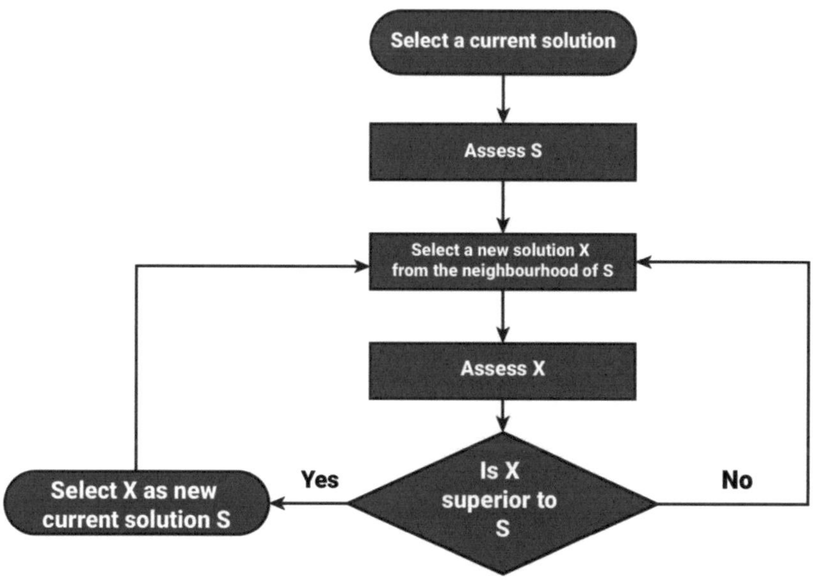

Figure 3.
Hill climbing.

2.2 Ant colony optimization

In nature-inspired algorithms the "ant colony optimization algorithm (ACO)" is a probabilistic intends to determine computational issues which can be diminished to finding the correct routes through graphs. Artificial Ants remain for multi-agent strategies motivated by the conduct of genuine ants. The pheromone-based communication of organic ants is frequently the dominating worldview utilized. Mixes of artificial ants and nearby local search algorithms have turned into a strategy for decision for various improvement tasks including some like vehicle routing and Internet routing [30].

Load balancing technique is essentially procedure which is used to find the optimal solution to given resource distribution problem in the given cloud computing scenario. Ant colony optimization considers the given problem as a graph and tries to find the optimal path and optimal resource allocation for the given problem can be found by using this specific procedure. The algorithm starts from randomness to optimization and alleviates the problem of the cluster. Initially, ants wander randomly and come back to the source by laying pheromone in their path. Then remaining ants follow the path set by the ants using pheromone [31]. There is a problem of pheromone getting evaporated this will allow for the shortest path determination. If it takes more time in this path, then the pheromone will evaporate there only shortest path pheromone survives to contribute to the shortest path. All the ants follow the same path to follow others eventually Thus after getting the shortest path whole ants to follow this path leading to the whole system following this path. This algorithm is more advantageous as it tackles the dynamic allocation problem easily. To solve the load balancing in a cloud environment the ant-based control system was designed. Each and every node was configured with capacity of being a destination, the probability of being destination, pheromone table. There are many variants for this technique namely Elitist depicts that global best solution gives pheromone update after every iteration, Max-Min ant system takes a min-max data structure and updates its value from minimum to maximum, Rank-based ant system takes all the possible solution and ranks according to sum of weights, continuous orthogonal ant colony takes additional angle parameter which makes for efficient searching, Recursive ant colony optimization takes solution and uses genetics to find out best solution [32].

The whole algorithm can be divided into different phases Edge selection and pheromone update. Initially randomly all ants are placed in the random order and after all the ants placed we updated pheromone level by using this formula Pxy<-(1-p)*pxy+ sum of all pheromone levels (**Figure 4**).

Here pxy is pheromone level which can be calculated by using this formula.

Del(the k) = J here j is the perimeter of curve xy for straight line use 0.

2.3 Artificial bee colony

The "artificial bee colony (ABO)" algorithm is a swarm based meta-heuristic algorithm. The algorithm is constructed unequivocally with respect to the model that is proposed for the scrounging conduct of honey bee settlements. The excellent comprises of three imperative parts: used and unemployed foraging honey bees, and sustenance sources. The underlying two sections, used and unemployed foraging honey bees, examine for rich sustenance sources, which is the third fragment, close to their hive. The model in like manner describes two driving techniques for lead which are basic for self-sorting out and total knowledge: enlistment of foragers to bounteous sustenance sources achieving positive information and surrender of poor

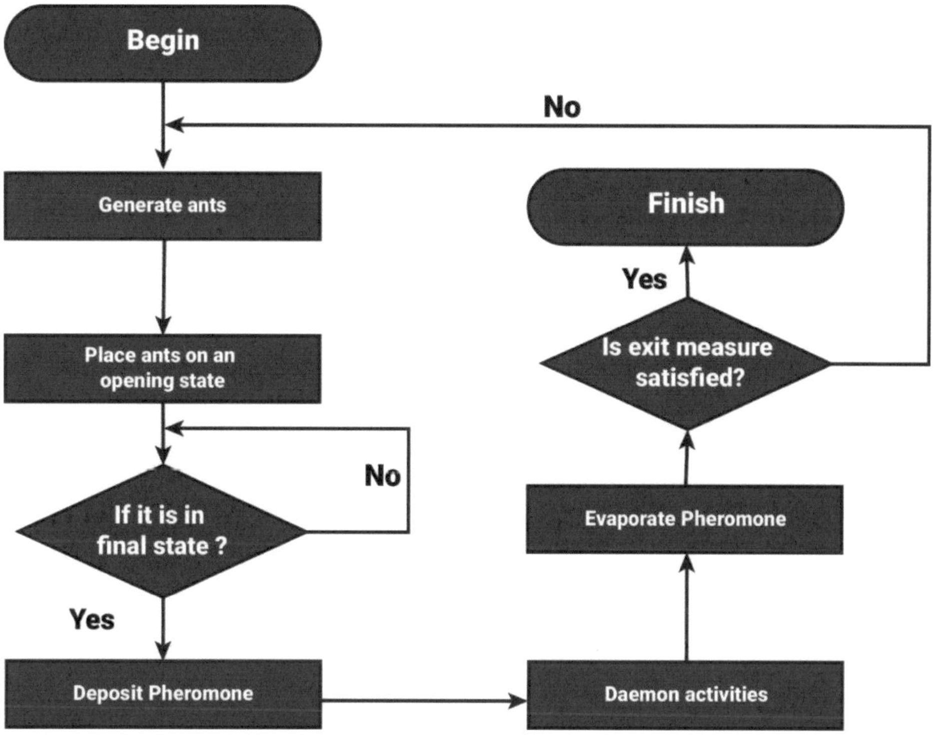

Figure 4.
Ant colony optimization.

sources by foragers causing negative feedback [33]. In ABC, a state of artificial forager honey bees check for rich phony sustenance sources. To apply ABC, the treated change issue is first changed to the issue of finding the best parameter vector which lessens a goal work. By then, artificial forager honey bees erratically discover a people of beginning arrangement vectors and a while later iteratively improve them by using the techniques: moving toward better arrangements utilizing a neighbor search mechanism while forsaking poor arrangements [34].

A technique called artificial bee colony optimization which is used to find the optimal solution to the given problem. It models the given problem as a graph and algorithm mimic the behavior of bees to approach a solution to the given Here problem refers to optimal resource allocation in given load balancer then resource allocation is done accordingly. ABC is an algorithm which takes inspiration from bees and tries to find the shortest path for the given graph cluster. Derviskaraboga developed it in 2005. The position of bees is modified to find out best position with the highest nectar since it is a population-based search procedure. Generally, bees perform a waggle dance to convey information regarding distances and directions. The whole graph system can be modeled into two significant components of food sources and foragers. Further classified of foragers can be done as unemployed foragers, employed foragers and experienced foragers. The algorithm can again be divided into four different phases [35].

(1) Initialization phase: the initial food sources are allocated randomly by using this formula.

$$\mathrm{Km} = \mathrm{lb} + \mathrm{random}(0,1) * (\mathrm{ub} - \mathrm{lb}) \qquad (1)$$

Here lb, ub are **lower** and upper bound of solution space of objective function.

(2) Employed bee phase: the neighboring food source Nmi is determined by using the following formula.

$$\text{Nmi} = \text{Kmi} + \text{random}[-1, 1] * (\text{Kmi} - \text{Kki}) \tag{2}$$

Here i is any **randomly** selected parameter index.

Here fitness is **calculated** by using the following formula and a greedy algorithm is applied.

$$\text{Fitness}(\text{Km}) = 1/1 + \text{obj}(\text{Km}),\ \text{obj}(\text{Km}) > 0,\ \text{Fitness}(\text{Km}) = 1 + |\text{obj}(\text{Km})|,\ \text{obj}(\text{Km}) < 0$$

Here **obj**(Km) is objective function of Km.

(3) Onlooker bee phase: the quantity of food sources is the ratio of total fitness to the individual bee fitness.

$$\text{Profitm} = \text{individual fitness/sum of all fitness.}$$

Onlooker bee uses the formula (2) for neighboring food source.

(4) Scout phase: the new solutions are randomly searched by the scout bees.

The new **solution** Km is randomly searched by following formula.

$$\text{Km} = \text{lb} + \text{random}(0, 1) * (\text{ub} - \text{lb})$$

Here ub and **lb** are upper and lower bounds to the solution phase (**Figure 5**).

2.4 Genetic algorithm

The approval of GA is situated on four considerations: populace estimate, mutation rate, crossover rate and the number of generations. To control the first rate individual among a masses, qualities of comparing people are ordered against the objective function. The formative structure through which another successor is conveyed is crossover and mutation. In the crossover mechanism, an offspring is conveyed by joining the characteristics of consolidating the qualities of chose people among populace while transformation causes some irregular changes in qualities of an individual in this manner delivering new hereditary person. The transformative mechanism is finished to the moment that joining criteria are satisfied [36].

The genetic algorithm is used to find out the optimal solution to the given cluster. The approval of GA is situated on four contemplations: populace estimate, A genetic algorithm is a refinement algorithm natural selection (survival of the fittest) is the one inspired genetic algorithm, Essentially we have to show the given bunch and attempt to explain it by utilizing a hereditary calculation. The genetic algorithm performs optimization and searching using three different bio-inspired operators such as mutation, crossover and selection [37]. One genetic representation of all the possible solutions and a fitness function is required by typical GA to figure out the quality of the given solution. The given problem can be either modeled as a graph or tree, and a genetic algorithm can be applied to this to determine the optimal solution. The genetic algorithm borrows its properties from natural selection by Darwin or Darwinian evolution theory. Hence we need to

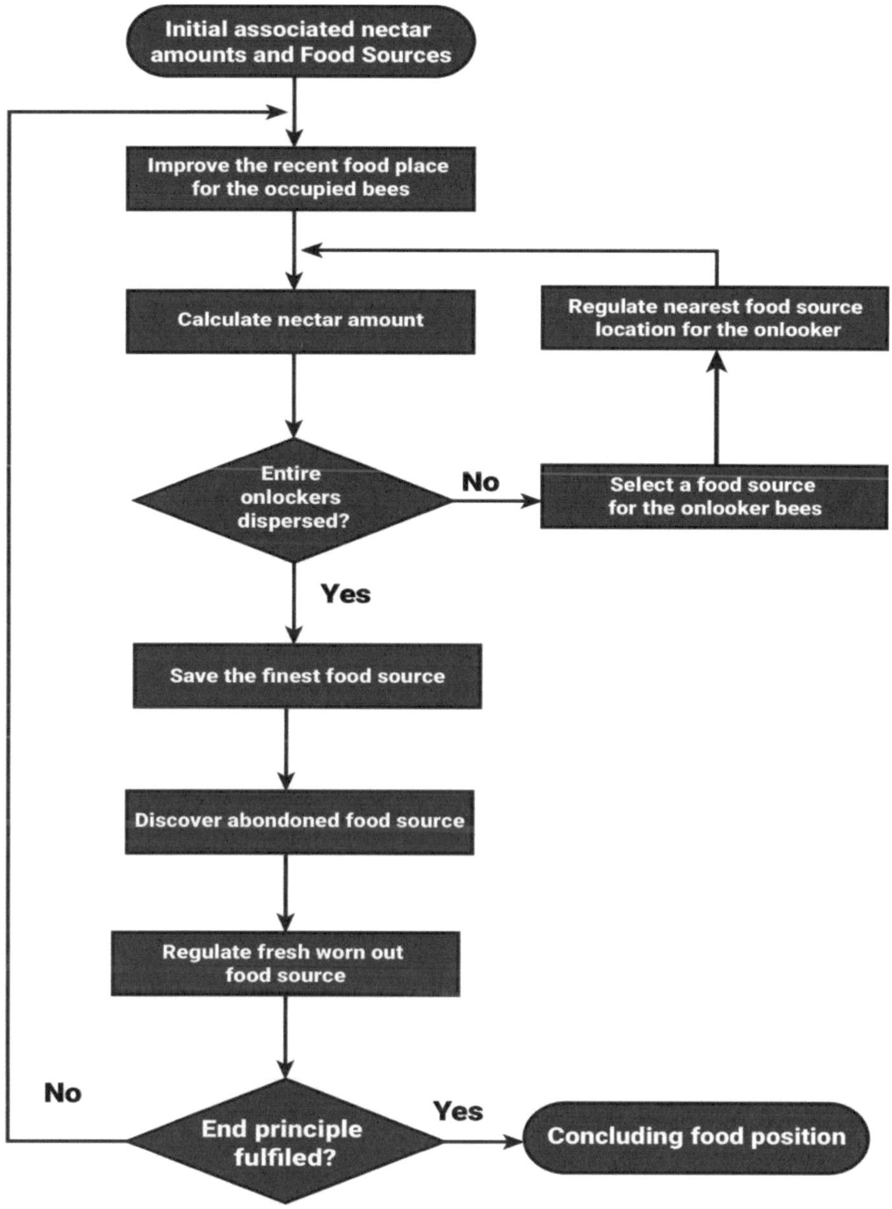

Figure 5.
Artificial bee colony (ABC).

implement these characteristics in our algorithm. The three significant steps in this algorithm are Initialization, selection and validation.

1. **Initialization step**: it is performed in randomness here we take all the possible solutions and list them out.

2. **Selection step**: with the help of probabilities possible solutions are listed out based on probabilities and parents are selected according to highest probabilities to form a solution.

3. **Termination step**: in the final step, we use different techniques like mutation, crossover, etc. to form a solution and this solution is evaluated by using fitness

function. This process is iterated until we form an optimal solution [38]. As it is an iterative technique this can perform for the infinite amount of time this should be terminated at some of the other instants to consider the optimal solution to the given problem. This can be terminated at either any of these cases:

a. If the latest solution persuades minimum criteria.

b. Fixed no of generations reached.

c. Allocated budget used up.

d. Manual inspection.

In its heart, genetic algorithm follows each of this basic mechanism:

a. Variation: this is implemented in the initialization part here we need to take variation as this would contribute to the formation of a solution.

b. Selection: this is implemented in the selection part here we take or consider the solution which has a higher probability to form a solution and remaining are eliminated.

c. Hereditary: this is implemented in the final stage whereby use of process like mutation allows the new solution to have two properties (**Figure 6**).

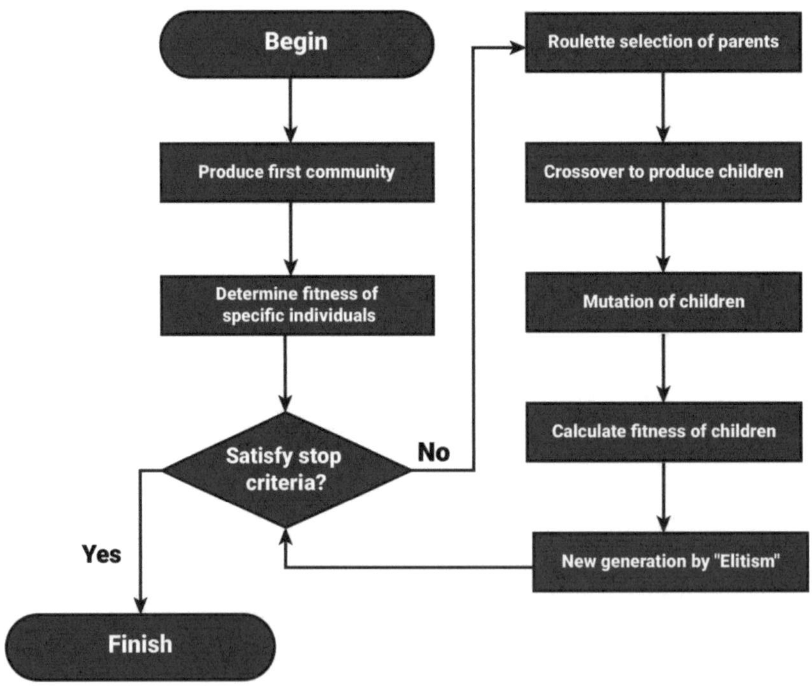

Figure 6.
Genetic algorithm.

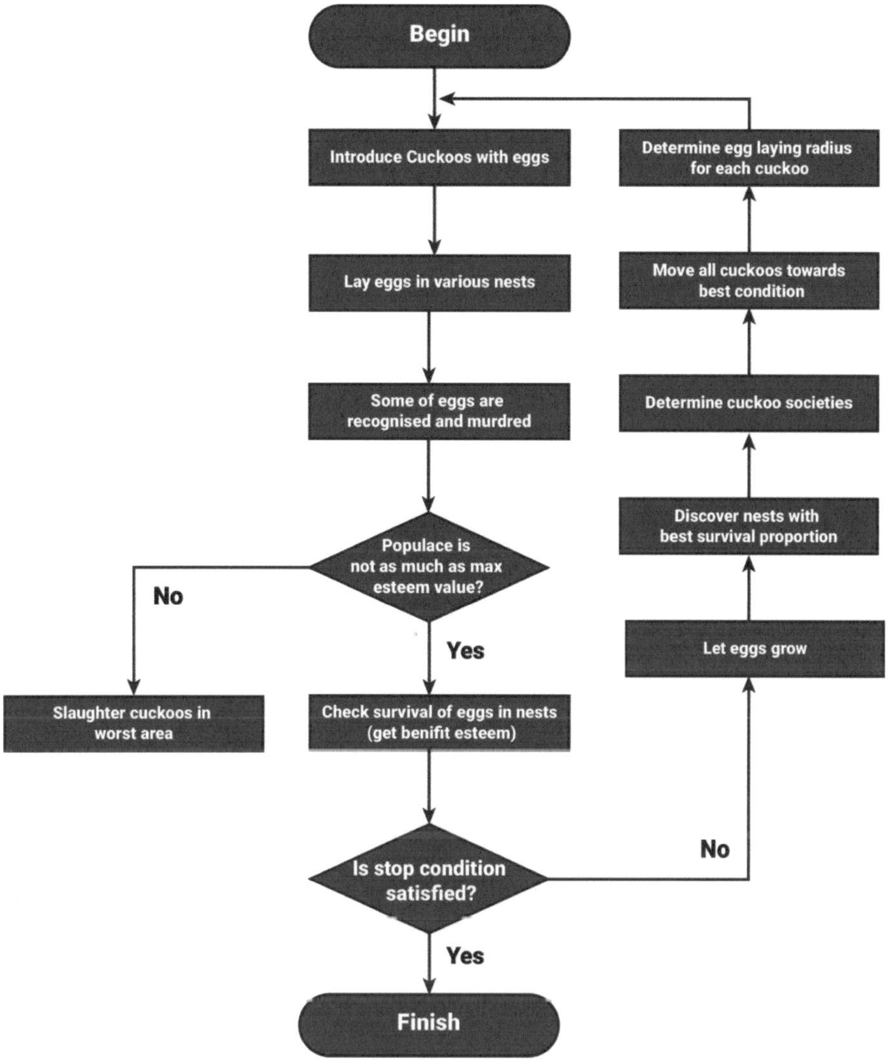

Figure 7.
Cuckoo search.

2.5 Cuckoo search

"Cuckoo search algorithm (CSA)" is another algorithm and is inspired by the raising behavior of the cuckoo bird they select their home by self-assertively accepting control over the home of some extraordinary winged animals for an age. They lay their eggs in the picked home of host cuckoo bird and drop the host winged animal's egg. The host bird either drop cuckoo fledgling's egg or surrender the whole home Some female cuckoo can copy their eggs like host fowl's egg and lay their eggs just before the laying of host feathered creature's egg. This grows the probability of their chick survival. Each egg in settle speaks to one arrangement, and the cuckoo flying creature's egg speaks to another arrangement. Wellness for every arrangement is handled, and settle with the high bore of eggs addresses the best game plan. The methodology is continued with aside from if a worldwide ideal arrangement is refined [39] (**Figure** 7).

To solve the problem of clusterin load balancer cuckoo search algorithm can be used. It does so by finding out the optimal resource allocation. Initially, we model the given cluster into a graph and then find the optimal resource allocation to this graph. Global solution to the given problem can be found using this technique. It essentially mimics the behavior of cuckoos and their behavior in laying eggs [40].

a. The algorithm is roughly based on these ideas.

b. How cuckoos lay their eggs in host nests.

c. How the eggs are hatched by hosts, if not detected and destroyed.

d. How can we mimic this behavior to make an algorithm.

This algorithm can be divided into five different steps:
Generating initial population:
We are generating host nests for k nests.

$$(l1, b1), (l2, b2), \text{ ---------- } (ln, bn)$$

These are optimal parameters

(1) Lay the cuckoo eggs (lk’, bk’) in the n nest.

n nest is randomly selected. Both have similar structure.

$$Lk' = lk + \text{Randomwalk}(\text{Levy flight})lk$$
$$bk' = bk + \text{Randomwalk}(\text{Levy flight})bk$$

(2) Compare the fitness of both eggs

The fitness can be found out by any statistical technique.

(3) Now replace the eggs according to the fitness value.

(4) If the host bird notices then leave that nest and search for other nest (to avoid local optimization).

(5) Repeat step 2–5 until we satisfy termination condition.

2.6 Firefly optimization

“Firefly algorithm (FA)” is the most heuristic algorithm for worldwide improvement, which is enlivened by the blazing behavior of firefly creepy crawlies. Xin-She Yang proposed this algorithm in 2008. The basic role of an (FA) is to go about as a flag framework to draw in different fireflies. Xin-She Yang defined this firefly algorithm by accepting that whole firefly are epicene with the goal that any single firefly will be pulled in to every other firefly. Engaging quality is relative to their shine, and for any two fireflies, the less brilliant one will be pulled in by the brighter one. In any case, the power diminishes as their aggregate length raises. If near are no fireflies luminous than an accustomed firefly, it will move randomly. The illumination should be related to the objective function [41] (**Figure 8**).

A load balancer is a device which is used to solve cluster problems in resource allocation. To solve the optimal resource allocation problem in the cluster we use

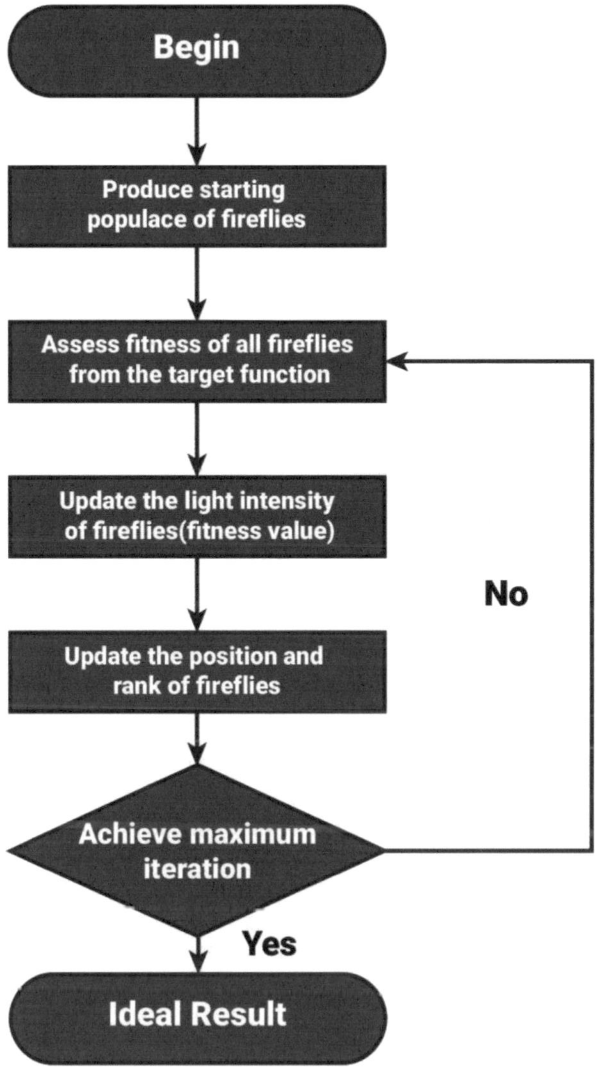

Figure 8.
Firefly optimization.

different kinds of algorithms. Here we are proposing this algorithm to solve the cluster of resource allocation in the load balancer. To solve the optimal resource allocation problem in the load balancer we use firefly optimization. Firefly optimization is unique in its approach as it tends to lead toward an optimal global solution. Essentially it mimics the behavior of fireflies, more explicitly flashing patterns of fireflies. Initially, we model the given problem into the graph, and we implement the firefly algorithm on this graph to find the optimal resource allocation [42].

1. Uses of flashing patterns like: communication, attracting prey, warning mechanism (female flies react toward the male's unique pattern of flashing in the same species).

2. Rules for thc algorithm:
   Rule 1: fireflies are unisex.

Rule 2: attraction increases as the brightness of light increases and decreases as the distance increases.
Rule 3: the brightness of fireflies is generally determined by the objective function.

3. The principle of the algorithm:

This algorithm can be divided into six different steps,
(a) Initializing the objective function:

$$I = I\_0\, e^{\left(-kd^2\right)}$$

Here k is the absorption coefficient and d is the distance.
As we know,

$$I(d) = I(s)/d^2$$

(b) Generating initial firefly population: we use this equation to initialize the firefly population

$$Mt + 1 = Mt + Bo * \left(e^{-}k\left(d^2\right)\right) + a * e$$

second term attraction third term randomization.
(c) Determine the intensity of light: find brightness for every firefly using objective function equation.
(d) Calculate the attractiveness of Firefly:

$$B = B0 * e^{\left(-kd^2\right)}$$

(e) Move the less bright fireflies to bright ones.
(f) Rank the flies and find the current best, Update the intensities.

2.7 Simulated annealing

For approximating the worldwide ideal of an inclined function, "simulated annealing (SA)" is a valuable method. It is often worn when the pursuit space is detached. For issues where revelation a close global optimum is broader than completing up an obvious local optimum in a shot volume of age, simulated annealing may be attractive over decisions, for example, gradient descent [43].

Simulated annealing is a strategy for approximating the worldwide ideal of a given function. It is utilized for global enhancement in expansive search space. It is utilized when search space is discrete. Annealing technology in metallurgy rouses it. A procedure including both warming and simultaneously cooling to increase the size and to decrease abandons in the given material. This procedure is utilized to locate a right arrangement. A load balancer is an apparatus which is utilized to find the optimal asset allotment of a given cluster issue in the cloud, and we can utilize a few sorts of metaheuristics to locate an optimal solution here we utilize recreated toughening to locate the optimal solution for the given issue. This solution begins by modeling graph of the given cluster, and we have to apply this strategy on the chart which will correspond to the optimal resource allocation solution in the cluster [44].

Initial slow cooling can be explored as the probability to accept the worst solution when this solution is running iteratively we get the best possible solution. After generating each solution the algorithm checks for its fitness and compares with previous one and picks the best one [45] (**Figure 9**).

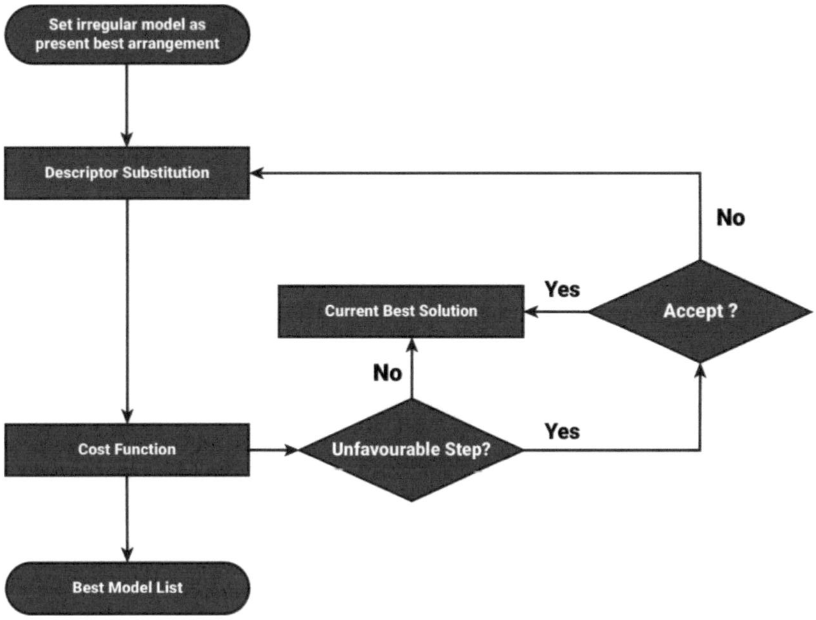

Figure 9.
Simulated annealing.

A process to find the best solution:

1. Selecting parameters: initially, we need to specify the following parameters in order to find an optimal solution:

    1. The space state.

    2. The energy goal function. E()

    3. The candidate generator neighbor()

    4. The acceptance probability p()

    5. The annealing scheduling temperature()

    6. Initial temperature()

2. Transition probability

3. Acceptance probability

4. Efficient candidate generation

5. Avoiding barrier

6. Cooling procedure

Each step can be mapped to the original simulated annealing process where we perform all these steps to get metal without defects here we use a similar procedure to find out the optimal solution.

Here we consider graph as metal and perform these tasks to find out the optimal solution.

2.8 Shuffled frog leaping algorithm

Load balancing using improved shuffled frog leaping algorithm. The "shuffled frog leaping algorithm (SFLA)" is a populace based algorithm excited by regular ridiculous. The virtual frogs go about as hosts or transporters of images where an image is a unit of social headway. The algorithm plays out an independent local search in each memeplex at the same time [46]. The local search is finished utilizing a particle swarm enhancement like technique adjusted for discrete issues, however, emphasizing a local search. To guarantee global investigation, the virtual frogs are intermittently rearranged and redesigned into new memeplexes in an approach indistinguishable from that well used in the rearranged complex evolution algorithm [47].

Load balancing technique essentially requires optimization technique to solve the given cluster problem of resource allocation. Any number of techniques can do this essentially we are using improved shuffled frog leaping algorithm which is also a part of swarm intelligence. To find the optimal solution to the given problem this algorithm imitates the behavior of frogs leaping and used this procedure. For load balancing with this technique, we require graph modeling of the cluster we get from

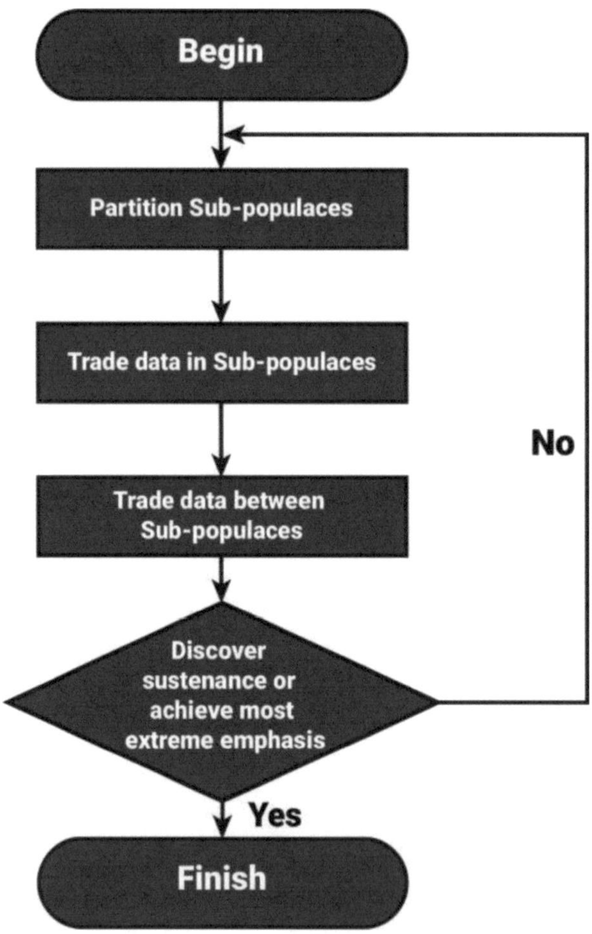

Figure 10.
Shuffled frog leaping algorithm.

the load balancer and then use this algorithm to find the optimal resource allocation. Every node in this algorithm consists of the capacity of VM, probability, etc. [48]. After graph modeling is done we continue with the stage of implementation using a frog leaping algorithm (**Figure 10**).

Essentially this algorithm can be classified into five phases:

1. Initialization phase: firstly, we need to declare the population size (k), no of subarrays (M), no of iterations in local exploration (u), no of algorithmic iterations (I), after that we need to generate no of frogs (k) randomly.

2. Fitness phase: firstly, we need to check the fitness of every frog using fitness functions.

3. Fitness functions: fitness = (1-(avg(load)/(avg(load)-least lode) + ((no of underloaded and overloaded nodes)/total no of nodes). Then after calculating the fitness of every frog, we need to sort them based on descending order of fitness values.

4. Formation of sub array and subarray: initially partitioning the sorted fitness array into subarrays as declared initially and now partitioning each subarray into another sub-array.

5. Local search phase: now perform a local search for every sub-array here is where this algorithm mimics frog leaping and perform search accordingly.

6. Convergence checking: now check for converging if convergence is satisfied this is optimal solution else repeat from frog fitness phase.

7. Off-spring generation phase: now perform this phase using fitness algorithm. Compare frogs and exchange information between sb, sw and continue this divide and conquer approach for u times, Now replace this final output by sb value.

The fitness for new product generated by the local search is calculated and this is updated solutions and this process continues until termination is given.

1. Termination phase:

    a. Number of generations.

    b. The fitness of the optimal solution.

    c. Time took.

2.9 Bat algorithm

The "bat algorithm (BA)" is the most eristic algorithm for global improvement. It was propelled by the allotment conduct of microbats, with fluctuating heartbeat rates of outflow and din. Xin-She Yang built up this algorithm in 2010. Each virtual bat flies heedlessly with a speed at a circumstance with a shifting recurrence or wavelength and tumult as it ventures and finds its prey, it changes recurrence, din and heartbeat surge rate. A local random walk escalates seek. Decision of the best continues to the point that specific stop criteria are met [49].

To solve the resource allocation problem in the given cluster a load balancer is an instrument which is utilized. We have to locate the ideal asset designation to take care of the issue of asset portion. The information for ideal asset portion is normally present in the cluster. We have to utilize any methods to locate the ideal answer for the given asset allotment issue. Here to locate the ideal answer for the given issue in the cluster we utilize the bat algorithm which impersonates the conduct of the bat. We need to use graph modeling of the given cluster to find the optimal solution initially. Essentially every node of cluster contains information about resource allocation, and when we use optimization on that graph, we find the optimal solution to the task which turns out to be optimal resource allocation for the given problem in the cluster. Here we are using Bat algorithm which is kind of swarm intelligence as it mimics the behavior of any flock of animals here we observe that we are mimicking the bat to find optimal solution hence bat algorithm [50].

Xin-She Yang proposed bat algorithm in the year 2010. It mimics the echolocation phenomenon seen in the bats. It uses sonar effect to navigate and communicate with other bats. Sonar is a form of ultrasonic sound only a few creatures like bats can understand and navigate through this. They calculate time delay to navigate, and the time delay is between emission and reflection. While reflecting bats use two notions zero means no emission one means maximum emission (**Figure 11**).

Ground Rules for this algorithm:

a. Bats can sense all kind of obstacles by this sound.

b. They fly randomly.

c. Loudness is always positive

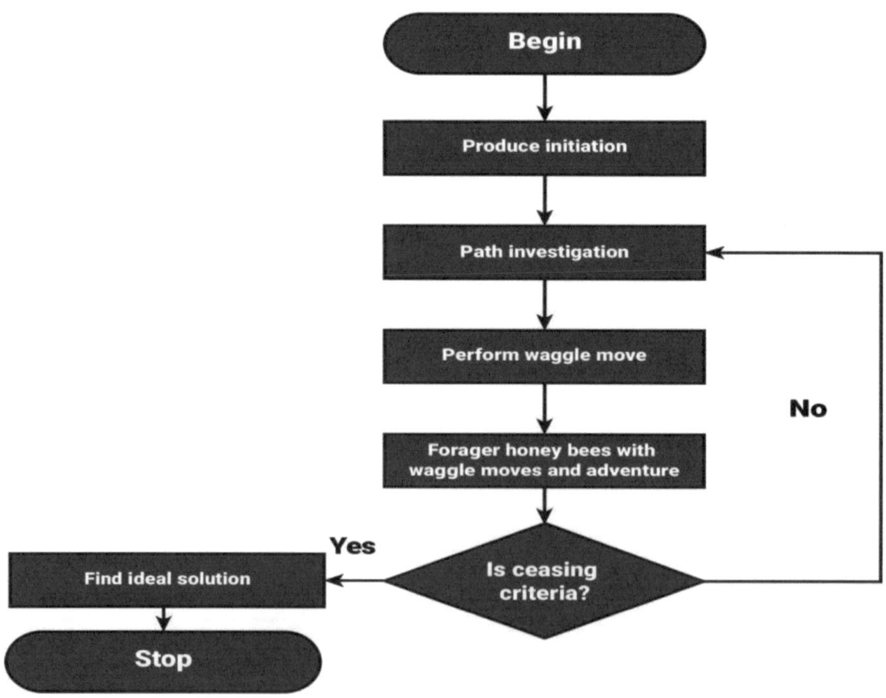

Figure 11.
Bat algorithm.

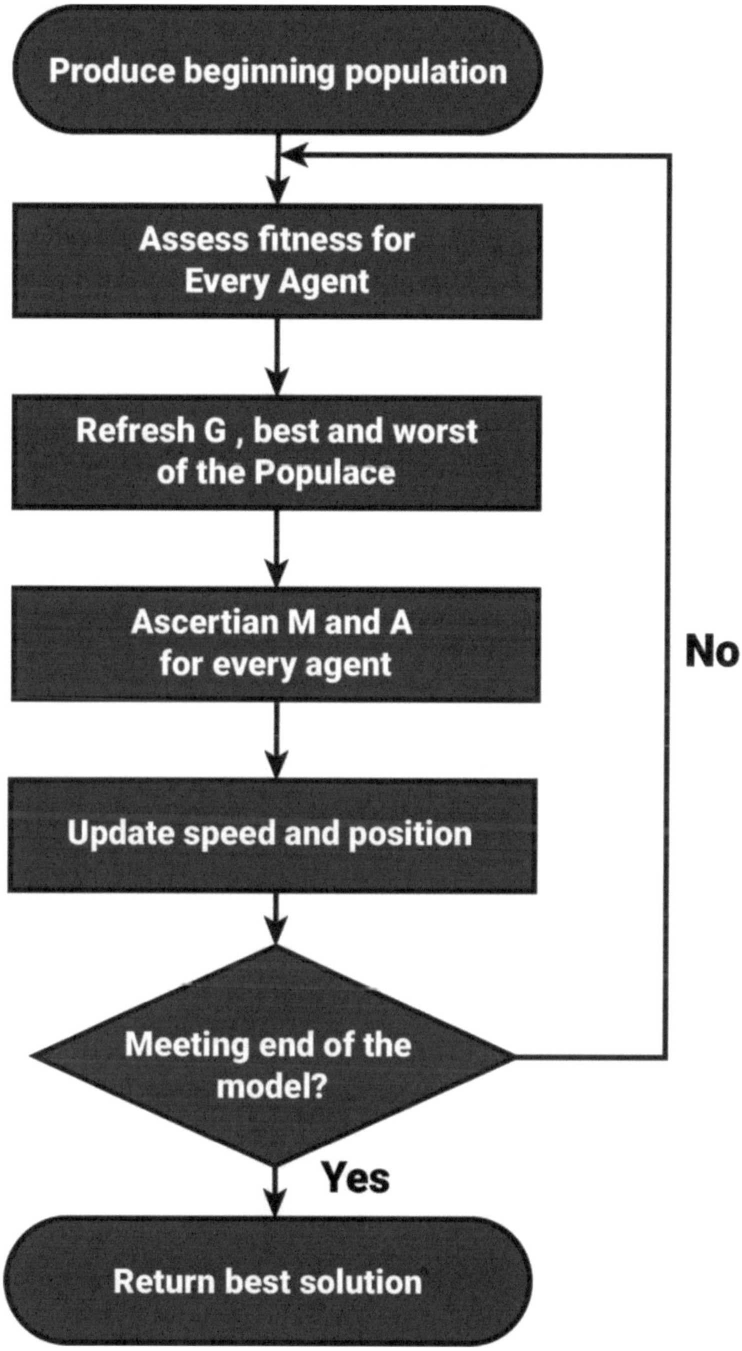

Figure 12.
Gravitational search algorithm.

Mathematical equations:

$$Fi = f0 + (f1 - f0) * K$$

$$Vit = vit - 1 + (xit - x*)fi$$

$$Xit = xit - 1 + vit$$

K is random value between 0 and 1.
x^* is current best solution.

Algorithm	Advantages	Disadvantages	Utilization	Adaptive
1. Hill Climbing	1. Better optimization technique rather than FIFO, DFS, etc. 2. Can be used with less computational resources. 3. Takes lesser time. 4. Can be used in Dynamic allocation.	1. Co-ordinate descent cannot be used as it takes huge amount of time to determine optimal solution. Stochastic hill climbing cannot be used as it does not find the exact optimal solution. Is less efficient when compared to other iterative algorithms like ant colony based, Genetic algorithm.	1. Can be used to find out initial optimal solution and can be put in iterative optimal solutions to find best solution.	1. Yes, because there are two different variants of techniques with respect to the given problem.
1. Ant colony Optimization	1. Best optimal solution can be found out for both static and dynamic problems. Takes relatively less time than traditional algorithms. As it has many variants like recursive one it can be used based on context.	1. Best optimal solution can be found but is relatively less effective than genetic and such kind of algorithms.	1. This can be used to both find initial solution and better the solution by using iterative variant.	1. Yes, it is adaptable as it has many variants
1. Artificial bee colony:	1. Can solve any kind of optimization problem in this case can solve any kind of resource allocation. Simple, Flexible and robust. Ability to explore local solutions. Ease of implementation.	1. The solution we get is optimal but not perfect solution. Not adaptable because every problem cannot be modeled into a graph. Cannot tackle dynamic allocation. Takes up more amount of time. Uses up more computational resources.	1. Static environment, Initial solution	1. No, because not every problem can be modeled as graph.
1. Genetic Algorithm	1. Takes up less amount of time. More desirable than all the traditional algorithms in terms of both time taken and forming output.	1. Needs more computational assets to produce the ideal answer for the issue. The algorithm irrespective of its accuracy can find difficulty to find global maximum and sometimes struck at local maxima. The termination is sometimes unclear that is optimal solution is always comparative. Cannot handle dynamic allocation.	1. Generally used for static problem and can be used for iterative solution making.	1. Yes, this can be used on any kind of problem cluster.

Algorithm	Advantages	Disadvantages	Utilization	Adaptive
1. Cuckoo Search	1. Deals with multi criteria optimization problem. Easy to implement. Simple to understand. Aims to speed up convergence.	1. Cannot tackle dynamic resource allocation.	1. Similar to ABC it can be utilized if we compromise on the quality of solution.	1. It is not adaptive as it does not tackle dynamic resource allocation.
1. Firefly optimization:	1. It can deal with highly non-linear problems. 2. It does not use velocities. 3. Does not require good initial start for optimization.	1. Global searching. 2. Slow converging speed. 3. High possibility to get trapped in the local optimum.	1. It can be used to find the accurate solution.	1. Not adaptive as it has slow converging speed.
1. Simulated annealing:	1. It can deal with inconsistent and noisy data. 2. To approach global optimum. 3. It is versatile as it does not rely on restrictive property of model.	1. Lot of choices are required to make it into actual algorithm. 2. It takes lot of computation time	1. It can be used to find accurate solution as it approaches to find the global solution.	1. Not adaptable as it takes lot of computational time.
1. Shuffled Frog Leaping Algorithm	1. Need not to model the given cluster as a graph. 2. Studies show that this algorithm is directing toward global optimal solution. 3. Optimal solution is found out in iterative manner.	1. Takes up more computational resources. 2. Takes up more amount of time.	1. This can be used for both static allocation problem and dynamic allocation problem.	1. Yes, as it does not always require graph modeling.
1. Bat Algorithm	1. Automatic zooming Parameter control Frequency tuning	1. Limited accuracy Unable to predict best solution.	1. Can be used to find the immediate solution.	1. Not adaptive as it has limited accuracy.
1. Gravitational search algorithm:	1. Less execution time. 2. Less computational resource. 3. More optimal solution.	1. This algorithm cannot be used on its own but can be used as support algorithm in hybrid functionality or for calculating fitness in other algorithms.	1. Utilization: for static allocation	1. No, it is not adaptive as it requires graph modeling.

Table 1.
Comparison of various algorithms.

$$Xnew = xold + H * at$$

H is a random value between −1 and 1 (Loudness decreases as the bat have found prey).

2.10 Gravitational search algorithm (GSA)

"Gravitational search algorithm (GSA)" is a reality discovering algorithm which is picking up enthusiasm among scientific community these days. Gravitational Search Algorithm (GSA) is a populace search algorithm based on Newton's law of gravity and the law of movement. GSA is represented to be more instinctual. In GSA, the specialist has four parameters which are the position, inertial mass, unique gravitational mass, and uninvolved gravitational mass. The arrangements in the GSA masses are called specialists, and they speak with each other through the gravity compel. Its mass measures the execution of these agents. Each agent is considered an object and all object move toward different items with all the more extensive mass because of gravity compel [17] (**Figure 12**).

To solve the cluster problem of resource allocation Load balancing is the technique used, and several techniques can be used to solve this clustering problem. One such way is using gravitational search algorithm. The gravitational search algorithm uses evolutionary computing [51].

Evolutionary computing is more efficient than traditional algorithms because the solutions do not stagnate in local minima and are faster and robust when compared to other different algorithms.

It is memoryless takes up less computational time than other algorithms. The gravitational search algorithm is a likeness to the particle swarm algorithm. Gravitational search algorithm was proposed by Rashedi et al. in the year 2009. This metaheuristic comes in the category of computational intelligence. Initially, a cluster from the load balancer is taken as an input, and this is modeled into a graph. To find out the optimal resource allocation now Gravitational search algorithm is applied to the graph. This algorithm is inspired from Newtonian mechanics and specifically from Newton's second law and law of gravitation [52].

3. Advantages and disadvantages of various algorithms

Advantages, disadvantages, utilization and adaptivity of various nature-inspired algorithms are listed in **Table 1**.

4. Conclusion

In this paper we chew over different basic concepts of cloud computing, the primary focus has been kept to understand load balancer and how it functions to do task allocation and different traditional algorithms which help in task allocation have also been discussed, and nature-inspired algorithms were discussed in specific. These were classified into different types like swarm based, genetically inspired, physics-based and different algorithms from each category were explained, and each was explained with different advantages, limitations, etc., and all were compared with respect to their positive points and working methodology and with relevance to load balancer An attempt has been made to survey out the different algorithms present in nature and provide relevant solution for optimal resource allocation for load balancer in cloud. All these different algorithms are nature-inspired and are used for global optimization. These algorithms are useful in finding optimized solutions for our lives by applying various laws and identifying the patterns of bats and flies like in Bat Algorithm, a search is identified by their local random walk and Firefly Algorithm.

5. Future scope

Different optimization techniques can be used as a hybrid to suit the appropriate usage to overcome the disadvantages of one over the other. New optimization techniques can be formulated from nature.

Author details

Surya Teja Marella* and Thummuru Gunasekhar
Department of Computer Science Engineering, Koneru Lakshmaiah Educational Foundation Vijayawada, Guntur, India

*Address all correspondence to: suryatejamarella@gmail.com

References

[1] Baran ME, Wu FF. Network reconfiguration in distribution systems for loss reduction and load balancing. IEEE Transactions on Power Delivery. 1989;**4**(2):1401-1407

[2] Randles M, Lamb D, Taleb-Bendiab A. A comparative study into distributed load balancing algorithms for cloud computing. In: 2010 IEEE 24th International Conference on Advanced Information Networking and Applications Workshops (WAINA); IEEE; April 2010. pp. 551-556

[3] Al Nuaimi K, Mohamed N, Al Nuaimi M, Al-Jaroodi J. A survey of load balancing in cloud computing: Challenges and algorithms. In: 2012 Second Symposium on Network Cloud Computing and Applications (NCCA); IEEE; December 2012. pp. 137-142

[4] Velayos H, Aleo V, Karlsson G. Load balancing in overlapping wireless LAN cells. In: 2004 IEEE International Conference on Communications; IEEE; June 2004; Vol. 7. pp. 3833-3836

[5] Tantawi AN, Towsley D. Optimal static load balancing in distributed computer systems. Journal of the ACM (JACM). 1985;**32**(2):445-465

[6] Yousaf FZ, Taleb T. Fine-grained resource-aware virtual network function management for 5G carrier cloud. IEEE Network. 2016;**30**(2): 110-115

[7] Cybenko G. Dynamic load balancing for distributed memory multiprocessors. Journal of Parallel and Distributed Computing. 1989;**7**(2):279-301

[8] Performance Tradeoffs in Static and Dynamic Load Balancing Strategies, NASA; March 1986

[9] Shirazi BA, Kavi KM, Hurson AR. Scheduling and Load Balancing in Parallel and Distributed Systems. IEEE Computer Society Press; 1995

[10] Cardellini V, Colajanni M, Yu PS. Dynamic load balancing on web-server systems. IEEE Internet Computing. 1999;**3**(3):28-39

[11] Schoonderwoerd R, Holland OE, Bruten JL, Rothkrantz LJ. Ant-based load balancing in telecommunications networks. Adaptive Behavior. 1997;**5**(2): 169-207

[12] Brendel J, Kring CJ, Liu Z, Marino CC. Resonate Inc. World-wide-web server with delayed resource-binding for resource-based load balancing on a distributed resource multi-node network. U.S. Patent 5, 774, 660. 1998

[13] Willebeek-LeMair MH, Reeves AP. Strategies for dynamic load balancing on highly parallel computers. IEEE Transactions on Parallel and Distributed Systems. 1993;**4**(9):979-993

[14] Miyazaki T, Wada M, Kawahara H, Sato M, Baba H, Shimada S. Dynamic load at baseline can predict radiographic disease progression in medial compartment knee osteoarthritis. Annals of the Rheumatic Diseases. 2002; **61**(7):617-622

[15] Devine KD, Boman EG, Heaphy RT, Hendrickson BA, Teresco JD, Faik J, et al. New challenges in dynamic load balancing. Applied Numerical Mathematics. 2005;**52**(2-3):133-152

[16] Kim C, Kameda H. An algorithm for optimal static load balancing in distributed computer systems. IEEE Transactions on Computers. 1992;**41**(3):381-384

[17] Rashedi E, Nezamabadi-Pour H, Saryazdi S. GSA: A gravitational search algorithm. Information Sciences. 2009; **179**(13):2232-2248

[18] Zaki MJ, Li W, Parthasarathy S. Customized dynamic load balancing for a network of workstations. In: 1996, Proceedings of 5th IEEE International Symposium on High Performance Distributed Computing; IEEE; August 1996. pp. 282-291

[19] Raz Y, Scherr AL, EMC Corp. Dynamic load balancing. U.S. Patent 5, 860, 137. 1999

[20] Trigui H, Cuthill R, Kusyk RG. Reverb Networks. Dynamic load balancing. U.S. Patent 8, 498, 207. 2013

[21] Raz Y, Vishlitzky N, Alterescu B, EMC Corp. Dynamic load balancing. U. S. Patent 6, 173, 306. 2001

[22] Hendrickson B, Devine K. Dynamic load balancing in computational mechanics. Computer Methods in Applied Mechanics and Engineering. 2000;**184**(2-4):485-500

[23] Sharma S, Singh S, Sharma M. Performance analysis of load balancing algorithms. World Academy of Science, Engineering and Technology. 2008; **38**(3):269-272

[24] Boyan JA, Littman ML. Packet routing in dynamically changing networks: A reinforcement learning approach. In: Advances in Neural Information Processing Systems; 1994. pp. 671-678

[25] Yang XS. Nature-Inspired Metaheuristic Algorithms. Luniver press; 2010

[26] Fister I Jr, Yang XS, Fister I, Brest J, Fister D. A brief review of nature-inspired algorithms for optimization. 2013. arXiv preprint arXiv:1307.4186

[27] Zang H, Zhang S, Hapeshi K. A review of nature-inspired algorithms. Journal of Bionic Engineering. 2010; **7**(4):S232-S237

[28] Mitchell M, Holland JH, Forrest S. When will a genetic algorithm outperform hill climbing. In: Advances in Neural Information Processing Systems; 1994. pp. 51-58

[29] Tsamardinos I, Brown LE, Aliferis CF. The max-min hill-climbing Bayesian network structure learning algorithm. Machine Learning. 2006;**65**(1):31-78

[30] Al Salami NM. Ant colony optimization algorithm. UbiCC Journal. 2009;**4**(3):823-826

[31] Dorigo M. Ant colony optimization. Scholarpedia. 2007;**2**(3):1461

[32] Yaseen SG, Al-Slamy NM. Ant colony optimization. IJCSNS. 2008;**8**(6):351

[33] Karaboga D. Artificial bee colony algorithm. scholarpedia. 2010;**5**(3):6915

[34] Karaboga D, Akay B. A comparative study of artificial bee colony algorithm. Applied Mathematics and Computation. 2009;**214**(1):108-132

[35] Karaboga D, Basturk B. A powerful and efficient algorithm for numerical function optimization: Artificial bee colony (ABC) algorithm. Journal of Global Optimization. 2007;**39**(3): 459-471

[36] Deb K, Pratap A, Agarwal S, Meyarivan TAMT. A fast and elitist multiobjective genetic algorithm: NSGA-II. IEEE Transactions on Evolutionary Computation. 2002;**6**(2): 182-197

[37] Morris GM, Goodsell DS, Halliday RS, Huey R, Hart WE, Belew RK, et al. Automated docking using a Lamarckian genetic algorithm and an empirical binding free energy function. Journal of Computational Chemistry. 1998;**19**(14): 1639-1662

[38] Deb K, Agrawal S, Pratap A, Meyarivan T. A fast elitist

non-dominated sorting genetic algorithm for multi-objective optimization: NSGA-II. In: International Conference on Parallel Problem Solving From Nature; Berlin, Heidelberg: Springer; September 2000. pp. 849-858

[39] Gandomi AH, Yang XS, Alavi AH. Cuckoo search algorithm: A metaheuristic approach to solve structural optimization problems. Engineering with Computers. 2013; **29**(1):17-35

[40] Yildiz AR. Cuckoo search algorithm for the selection of optimal machining parameters in milling operations. The International Journal of Advanced Manufacturing Technology. 2013;**64** (1-4):55-61

[41] Yang XS. Firefly algorithm, stochastic test functions and design optimisation. International Journal of Bio-Inspired Computation. 2010;**2**(2): 78-84

[42] Yang XS. Firefly algorithm, levy flights and global optimization. In: Research and Development in Intelligent Systems XXVI. London: Springer; 2010. pp. 209-218

[43] Corana A, Marchesi M, Martini C, Ridella S. Minimizing multimodal functions of continuous variables with the "simulated annealing" algorithm corrigenda for this article is available here. ACM Transactions on Mathematical Software (TOMS). 1987; **13**(3):262-280

[44] Kirkpatrick S, Gelatt CD, Vecchi MP. Optimization by simulated annealing. Science. 1983;**220**(4598): 671-680

[45] Szu H, Hartley R. Fast simulated annealing. Physics Letters A. 1987;**122** (3-4):157-162

[46] Eusuff MM, Lansey KE. Optimization of water distribution network design using the shuffled frog leaping algorithm. Journal of Water Resources Planning and Management. 2003;**129**(3):210-225

[47] Eusuff M, Lansey K, Pasha F. Shuffled frog-leaping algorithm: A memetic meta-heuristic for discrete optimization. Engineering Optimization. 2006;**38**(2):129-154

[48] Rahimi-Vahed A, Mirzaei AH. A hybrid multi-objective shuffled frog-leaping algorithm for a mixed-model assembly line sequencing problem. Computers & Industrial Engineering. 2007;**53**(4):642-666

[49] Yang XS. A new metaheuristic bat-inspired algorithm. In: Nature Inspired Cooperative Strategies for Optimization (NICSO 2010). Berlin, Heidelberg: Springer; 2010. pp. 65-74

[50] Yang XS, Hossein Gandomi A. Bat algorithm: A novel approach for global engineering optimization. Engineering Computations. 2012;**29**(5):464-483

[51] Rashedi E, Nezamabadi-Pour H, Saryazdi S. BGSA: Binary gravitational search algorithm. Natural Computing. 2010;**9**(3):727-745

[52] Rashedi E, Nezamabadi-Pour H, Saryazdi S. Filter modeling using gravitational search algorithm. Engineering Applications of Artificial Intelligence. 2011;**24**(1):117-122

Section 3

Cloud Computing and Data Science: Exploring the Benefits of Task Scheduling on Such Environments

Chapter 6

Looking at Data Science through the Lens of Scheduling and Load Balancing

Diórgenes Eugênio da Silveira, Eduardo Souza dos Reis, Rodrigo Simon Bavaresco, Marcio Miguel Gomes, Cristiano André da Costa, Jorge Luis Victoria Barbosa, Rodolfo Stoffel Antunes, Alvaro Machado Júnior, Rodrigo Saad and Rodrigo da Rosa Righi

Abstract

The growth in data generated by private and public organizations leads to several opportunities to obtain valuable knowledge. In this scenario, data science becomes pertinent to define a structured methodology to extract valuable knowledge from raw data. It encompasses a heterogeneous group of techniques that challenge the implementation of a single platform capable of incorporating all the available resources. Thus, it is necessary to formulate a data science workflow based on different tools to extract knowledge from massive datasets. In this context, high-performance computing (HPC) provides the infrastructure required to optimize the processing time of data science workflows, which become a collection of tasks that must be efficiently scheduled to provide results in acceptable time intervals. While few studies explore the use of HPC for data science tasks, in the best of our knowledge, none conducts an in-depth analysis of scheduling and load balancing on such workflows. In this context, this chapter proposes an analysis of scheduling and load balancing from the perspective of data science scenarios. It presents concepts, environments, and tools to summarize the theoretical background required to define, assign, and execute data science workflows. Furthermore, we are also presenting new trends concerning the intersection of data science, scheduling, and load balance.

Keywords: scheduling, load balance, high-performance computing, data science, big data

1. Introduction

Private corporate networks, as well as the Internet, generate and share data at ever increasing rates. This unconstrained growth can easily lead disorganization and, as a consequence, missed opportunities to analyze and extract knowledge for these data. There is an essential difference between the concepts of data and information. Data cannot express something outside a particular field of expertise.

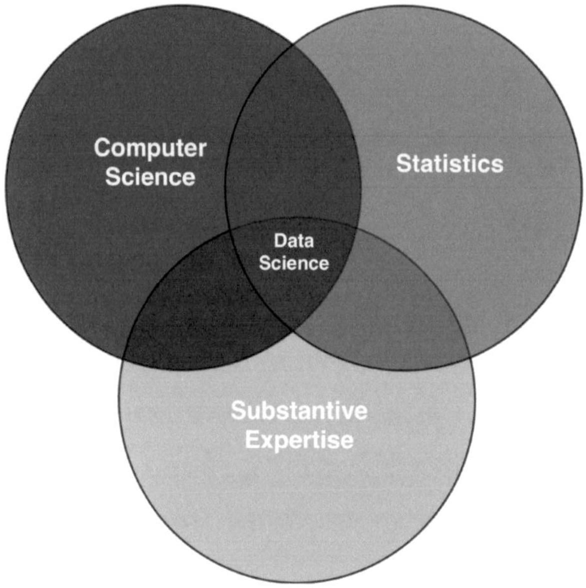

Figure 1.
Venn's diagram for correlating the influence of other research areas on data science.

In turn, information enables the coherent transmission of knowledge. Data science aims to close the gap between data and knowledge through the use of computational tools. More specifically, data science is a tool for converting raw data into knowledge [1]. The field of data science leverages many methods originating from computer science and statistics [2]. **Figure 1** illustrates a Venn's diagram that correlates the research areas with major influence in data science.

Although data science receives significant influence from expert knowledge, it is plausible to say that a data scientist knows more about computer science than a statistician and more about statistics than a computer scientist [3]. Besides, it also encompasses the intersection of data analytics and machine learning. Therefore, data science encompasses a heterogeneous group of studies and methodologies such as big data, machine learning, data analytics, and statistics, which challenge the implementation of a single platform capable of incorporating all the available techniques.

There are a variety of widely adopted platforms available for data analysis and knowledge extraction, for example, Tableau,[1] Dataiko,[2] Microsoft Azure Machine Learning Studio,[3] Orange BioLab,[4] each one suitable for a specific step of a data science process. A workflow can be formulated based on the coordinated application of different tools to extract knowledge from massive datasets. In this context, the use of cloud platforms for data science steadily grows because they offer scalability and distributed execution of individual tasks.

In data science, a large dataset allows the generation of a more in-depth model, which provides more robust insights because there are more instances to compose the statistical analysis of data. One of the most relevant aspects regarding a dataset is the quality of available data. Thus, before the use of any statistical method, the

[1] https://www.tableau.com/

[2] https://www.dataiku.com/

[3] https://azure.microsoft.com/en-us/services/machine-learning-studio/

[4] https://orange.biolab.si/

dataset must go through a cleaning process that ensures the uniformity of values and the elimination of duplicated data. On one hand, a large dataset with high-quality data enables an insightful model. On the other hand, the computational power required to process data is directly proportional to the size of the available dataset. In this scenario, high-performance computing (HPC) provides the infrastructure (clusters, grids, and cloud platforms) required to optimize the processing time of data science workflows. In particular, data science demands are transformed in a collection of tasks, with or without the notion of dependency among them, which must be efficiently scheduled along the computational resources [memory, processors, cores, cluster nodes, graphical processing unit (GPU) cores, grid nodes, and virtual machines, for example] to provide the results in an acceptable time interval. To map such tasks to resources, a scheduling policy takes place where load balancing algorithms are important to provide a better execution equilibrium among the tasks and a fast response time, mainly when considering either dynamic or heterogeneous environments. While some articles explore the use of HPC for data science tasks [4–6], in the best of our knowledge, there are no studies that conduct an in-depth analysis of how the aspects of scheduling and load balancing affect data science workflows.

Hence, the present book chapter proposes an analysis of scheduling and load balancing from the perspective of data science scenarios. Furthermore, it presents concepts, environments, and tools for data science to summarize the theoretical background required to understand the definition and execution of data science workflows. Even though its focus lies on presenting concepts, the chapter also illustrates new trends concerning the intersection of data science, scheduling, and load balance.

The remainder of this chapter is organized as follows: Section 2 presents an in-depth explanation of concepts, workflow, problem classes, and tools used by data science. Section 3 explores scheduling and load balancing as tools to leverage the computational power required by data science applications. Section 4 points to open challenges and trends in the use of HPC applied to data science problems. Finally, Section 5 concludes the chapter with closing remarks and directions for future work.

2. Fundamental concepts

This section presents the fundamental concepts related to data science. These are key to understand the concept of HPC, more specifically scheduling and load balancing, impact data science processes, as discussed later in the chapter. The remainder of the section discusses the fundamental components of a data science pipeline, as observed in real-world scenarios.

2.1 Data science workflow

Data science is highly dependent on its application domain and employs complex methods. Nevertheless, it has a very organized pipeline, which varies in the number of steps required to extract knowledge. Current work explores a pipeline that varies between five and seven steps, but in all cases, the process yields similar outputs. This section aims at presenting the most complete process, composed of seven steps, widely used by both companies and researchers. **Figure 2** depicts the flow of information step by step. Moreover, the seven proposed steps can be enumerated as:

1. business understanding;
2. data extraction;
3. data preparation;
4. data exploring;
5. data model;
6. results evaluation; and
7. implementation.

Step 1 refers to the process of understanding in which context the data are inserted on, and what is the expected output. This is a high time-consuming process in a project. However, data scientist must have a deep understanding about the application domain to validate the model's structure as well as its outputs. After understanding the scope of the project, on Step 2, exploring the data that correlate with the problem understood in the last step. These data can be hosted at the client or not. If the client does not have useful data available, the data scientist must look for a synthetic or publicly available dataset to extract the knowledge. Furthermore, on Step 3, techniques are employed to clean data because there is a high chance that it is unorganized or unreadable, so it is necessary to preprocess and standardize it. An example of this step is a dataset that has a column with country names, but in some registers, the value of this column is "Brasil" and in others, it is "Brazil," both values symbolize the same information but are encoded in different languages. Regarding Step 4, the data are organized, and it is indispensable to execute a

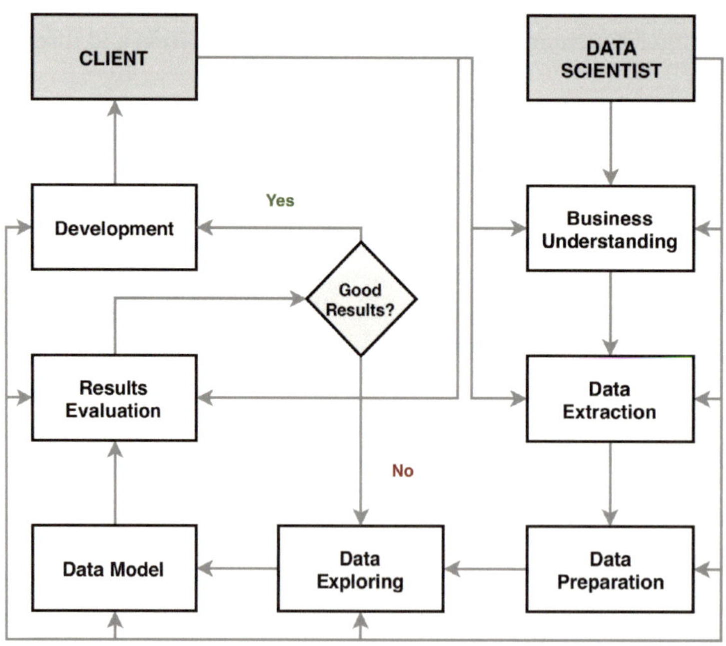

Figure 2.
Flow diagram of data science steps.

detailed analysis in order to figure out patterns or insights that would be valuable to the client. In this stage, the data scientist usually uses plotting techniques to make the data more readable and figure out information.

On Step 5, previously identified insights serve as input. But at this step, it is vital to fully understand the data since, without formal knowledge, it is very hard to fit a model that correctly represents it. At this stage, it is required that the data scientist uses computer science expertise to choose the better approach to plan and validate the model. In Step 6, outputs generated by the model are evaluated in order to analyze how useful they actually are. Usually, this evaluation is conducted by the client and the data scientist together, to examine graphs, numbers, and tables and define if the model generated acceptable results. Finally, on Step 7, the results are validated, and the model is ready to be implemented and deployed in production; thus, it is applied to prediction tasks, having real data as input. The architecture used to implement the model is very important, that is, it is necessary to understand what will be used for the client application since if the client needs a real-time response, the structure will be very different than a nonreal-time scenario.

2.2 Solving problems with data science

Data science is not exclusively employed in business scenarios, and it can be generalized to a plethora of applications, such as in Obama's campaign for US presidential elections in 2012. In the context of this election, technology was applied to identify who were the voters that should receive more attention and marketing influence. Some analysts highlighted the use of data science as fundamental to Obama's victory. However, data science is not limited to the analysis of scenarios, such as in the above example. Many other challenges can benefit from solutions based on data science methodologies. For instance, some problem classes in data science are pattern detection, anomaly detection, cleaning, alignment, classification, regression, knowledge base construction, and density estimation [7]. These classes of problems are explored next.

2.2.1 Pattern detection

The patterns existing in a dataset are not always easily identifiable due to the organization of the data. It is the method employed to discover information standards in the dataset. Figuring patterns hidden into data is a relevant task in several scenarios, for example, to join the clients with similar characteristics, such as those with the same taste or opinion.

2.2.2 Anomaly detection

The distribution of a dataset regards the positions of data points among each other in the dataset. Usually, the representation of this distribution employs a Cartesian plane, in which each point is an instance of the dataset. Within this representation, regions with a defined concentration of data points become clusters. Therefore, outliers are the data points that are too far from these clusters. Anomaly detection aims to classify each data into a dataset as an outlier or not. For example, the banks employ this approach in the scenario of fraud detection. In this scenario, transactions of a client become clusters, and a new transaction is considered as unclassified data.

2.2.3 Cleaning

Dataset contents can present a broad variety of formats, for example, dates, numeric values, and text data. In many cases, values may also be differently formatted across dataset entries, even for the same field, due to human error or lack of standardization. This situation incurs in errors in the dataset, resulting in possible information loss when processing it. The best solution to this problem is to employ methods that manage the dataset contents and standardize the data. For example, in a dataset containing clients birthdays, it is possible that a user fills the information with an invalid date. This case results in information loss because the analysis cannot extract useful information related to the user's birthday.

2.2.4 Alignment

An organized and standardized dataset is fundamental for the generation of trustworthy outputs from data science processes. The process of data alignment is an essential step in dataset standardization. It involves updating the dataset to avoid the use of multiple values to represent the same information. For example, in a dataset containing a gender field, users may use both "M" and "Male" values to represent the same gender. In this context, it is fundamental to unify both entries in one because the information in both is the same.

2.2.5 Classification

Classification regards assigning specific labels to entries in a dataset. A label is any information that presents a limited scope of possibilities, for example, a dozen options present in a specific dataset field. Examples of labels include sentiments, states, and the scale of integer numbers. This dataset field can then be used to group multiple dataset entries according to the unique values that the label may take. For example, in a dataset about client's purchases, each product may be associated with a set of keywords. These keywords can then be used to classify the types of purchases a particular client makes, enabling targeted recommendations for other products.

2.2.6 Regression

Datasets do not always contain fields with labels that enable the classification of data. Nevertheless, in some problems, it is necessary to label values without a restricted group of options, for example, using a field that contains real numbers. This scenario requires a regression approach to estimate which classes an entry should receive, without considering the limited options available in labels. For example, in a dataset with prices and sizes of houses in New York, it is possible to use regression to estimate the price of a new house according to its size. Although the regression problem has an output similar to classification, its output does not have a limited set of values, as occurs in labeling.

2.2.7 Knowledge base construction

Datasets are essential for data science, and the problem of knowledge base construction refers to the process of compiling information to create them. Frequently, this process requires the use of cleaning and alignment methods to

Stage	Tools
Store data	MySql, Mongo DB, Cassandra, PLSql, Redis, HBase
Data preparation	Apache Hive
Data exploring	Knime, Elasticsearch
Data model	Python, R, Julia, Clojure, SPSS, SAS, Apache Manhout
Results evaluation	Tableau, Cognos, ggplot, QlikView, Power BI
Development	Apache Hadoop, Java, Scala, C, Apache Spark, Haskell

Table 1.
Tools used per step of data science.

standardize data. There are a broad group of knowledge bases on the Internet, for example, Kaggle,[5] UCI Repository,[6] Quandl,[7] and MSCOCO.[8]

2.2.8 Density estimation

Density estimation focuses on identifying the clusters that group sets of data points that represent the entries in a dataset. This process is a fundamental step to generate the clusters required by anomaly detection methodologies, as described above. Clustering is another suitable technique to identify groups of entries that may contain related knowledge within a dataset.

2.3 Data science tools

It is difficult to find a tool that fits all data science processes because, as previously mentioned, there are multiple steps with a variety of methods available for use. Hence, there are specific tools for each step, which will provide the most appropriate result. **Table 1** summarizes the most commonly cited tools for each one of the data science workflow steps. The table has a row that is not considered a step of the process, but it is fundamental to results that are Storage Data, which refer to all technologies used for persisting the data in an environment. In the section of data model, some programming languages are cited, but it is hard for a data scientist to employ a language without libraries. For example, using Python is very usual to use libraries such as pandas, sci-kit learn, numpy, and ggplot.

3. Exploring scheduling and load balancing on data science demands

The scheduling problem, in a general view, comprises both a set of resources and a set of consumers [8]. Its focus is to find an appropriate policy to manage the use of resources by several consumers in order to optimize a particular performance metric chosen as a parameter. The evaluation of a scheduling proposal commonly considers two features: (1) performance and (2) efficiency [9]. More specifically, the evaluation comprises the obtained scheduling as well as the time spent to execute the scheduler policies. For example, if the parameter to analyze the

[5] https://www.kaggle.com
[6] https://archive.ics.uci.edu/ml/index.php
[7] https://www.quandl.com/
[8] http://cocodataset.org/home

achieved scheduling is the application execution time, the lower this value, the better the scheduler performance. In turn, efficiency refers to the policies adopted by the scheduler and can be evaluated using computational complexity functions [10].

The general scheduling problem is the unification of two terms in everyday use in the literature. There is often an implicit distinction between the terms scheduling and allocation. Nevertheless, it can be argued that these are merely alternative formulations of the same problem, with allocation posed in terms of resource allocation (from the resources point of view), and scheduling viewed from the consumers' point of view. In this sense, allocation and scheduling are merely two terms describing the same general mechanism but described from different viewpoints. One important issue when treating scheduling is the grain of the consumers [11]. For example, we can have a graph of tasks, a set of processes, and jobs that need resources to execute. In this context, scheduling schemes for multiprogrammed parallel systems can be viewed in two levels. In the first level, processors are allocated to a specific job. In the second level, processes from a job are scheduled using this pool of processors.

We define static scheduling considering the scheduling grain as a task [8]. If data such as information about the processors, the execution time of the tasks, the size of the data, the communication pattern, and the dependency relation among the tasks are known in advance, we can affirm that we have a static or deterministic scheduling model. In this approach, each executable image in the system has a static assignment to a particular set of processors. Scheduling decisions are made deterministically or probabilistically at compile time and remain constant during runtime. The static approach is simple to be implemented. However, it is pointed out that it has two significant disadvantages [11]. First, the workload distribution and the behavior of many applications cannot be predicted before program execution. Second, static scheduling assumes that the characteristics of the computing resources and communication network are known in advance and remain constant. Such an assumption may not be applied to grid environments, for instance.

In the general form of a static task scheduling problem, an application is represented by a directed acyclic graph (DAG) in which nodes represent application tasks, and edges represent intertask data dependencies [12].

Each node label shows computation cost (expected computation time) of the task, and each edge label shows intertask communication cost (expected communication time) between tasks. The objective function of this problem is to map tasks onto processors and order their executions, so that task-precedence requirements are satisfied, and the minimum overall completion time is obtained.

In the case that all information regarding the state of the system as well as the resource needs of a process is known, an optimal assignment can be proposed [11]. Even with all information required for the scheduling, the static method is often computationally expensive getting to the point of being infeasible. Thus, this fact results in suboptimal solutions. We have two general categories within the realm of suboptimal solutions for the scheduling problem: (1) approximate and (2) heuristic. Approximate scheduling uses the same methods used in the optimal one, but instead exploring all possible ideal solutions, it stops when a good one is achieved. Heuristic scheduling uses standard parameters and ideas that affect the behavior of the parallel system. For example, we can group processes with higher communication rate to the same local network or sort works and processors in lists following some predefined criteria in order to perform an efficient mapping among them (list scheduling).

Dynamic scheduling works with the idea that a little (or none) a priori knowledge about the needs and the behavior of the application is available [9]. It is also unknown in what environment the process will execute during its lifetime. The arrival of new tasks, the relation among them, and data about the target architecture are unpredictable, and the runtime environment takes the decision of the consumer-resource mapping. The responsibility of global scheduling can be assigned either to a single processor (physically nondistributed) or practiced by a set of processors (physically distributed). Within the realm of this last classification, the taxonomy may also distinguish between those mechanisms that involve cooperation between the distributed components (cooperative) and those in which the individual processors make decisions independent of the actions of the other processors (noncooperative). In the cooperative case, each processor has the responsibility to carry out its portion of the scheduling, but all processors are working toward a common system-wide goal.

Data science comprises the manipulation of a large set of data to extract knowledge [13, 14]. To accomplish this, we have input that is passed through processing engines to generate valuable outputs. In particular, this second step is usually processed as sequential programs that implement both artificial intelligence and statistical-based computational methods. We can take profit from the several processing cores that exist in today's processors to map this sequential demand to be executed in a multithreading program. To accomplish this, Pthreads library and OpenMP are the most common approaches to write multithread parallel programs, where each thread can be mapped to a different core, so exploiting the full power of a multiprocessor HPC architecture.

In addition to multiprocessor architectures, it is possible to transform a sequential code in message passing interface (MPI)-based parallel one, so targeting distributed architectures such as clusters and grids [15]. In this way, contrary to the prior alternative that encompasses the use of standard multiprocessor systems, the efficient use of MPI needs a parallel machine that generally has higher financial costs. Also, a distributed program is more error prone, since problems in the nodes or the network can put all application down. Repairing these future problems sometimes is not trivial, requiring graphical tools to observe processes' interactions. Finally, in addition to multicore and multicomputer architectures, we also have the use of GPU, where graphic cards present a set of nongeneral purpose processors to execute vector calculus much faster than the conventional general-purpose processors [14]. The challenge consists in transforming a sequential code in a parallel one in a transparent way at the user viewpoint, in such a way the data science demand can run faster in parallel deployments. Moreover, the combination of these three aforesaid parallel techniques is also a challenge since optimizations commonly vary from one application to another.

Cloud computing environments today also represent a viable solution to run data science demands [16]. Providers such as Amazon EC2, Microsoft Azure, and Google Cloud have HPC-driven architectures to exploit multiprocessor, multicomputer, and GPU parallelism. In particular, different from standard distributed systems, cloud computing presents the resource elasticity feature where an initial deployment can be on-the-fly changed following the input demand. Thus, it is possible to scale resources in or out (through the addition or removal of containers/virtual machines) or to scale down or up (by performing resource resizing in virtual units) in a transparent way to the user. Logically, the own data science application must be written in such a way to take profit of newly available resources as the current set of working resources.

The CPU load is the most common metric to drive resource elasticity data science demands since most of them execute CPU-bound artificial intelligence-based algorithms. Any network data manipulation through the TCP protocol uses CPU cycles since this is a software protocol executed in the kernel of the operating system and executes software routines to provide data transfer reliability.

Load balancing and resource scheduling are sometimes seen as having the same functionality. However, there is a slight difference: one of the members of resource scheduling is the scheduling, and this policy can employ or not load balancing algorithms [14]. The basic idea of load balancing is to attempt to balance the load on all processors in such a way to allow all processes on all nodes to proceed at approximately with the same rate. The most significant reason to launch the load balancing is the fact that exists an amount of processing power that is not used efficiently, mainly in dynamic and heterogeneous environments, including grids. In this context, schedulers' policies can use load balancing mechanisms for several purposes, such as: (1) to choose the amount of work to be sent to a process; (2) to move work from an overloaded processor to another that presents a light load; (3) to choose a place (node) to launch a new process from a parallel application; and (4) to decide about process migration. Load balancing is especially essential for some parallel applications that present synchronization points, in which the processes must execute together with the next step.

The most fundamental topic in load balancing consists of determining the measure of the load [13, 15]. There are many different possible measures of load including: (1) number of tasks in a queue; (2) CPU load average; (3) CPU utilization at specific moment; (4) I/O utilization; (5) amount of free CPU; (6) amount of free memory; (7) amount of communication among the processes; and so on. Besides this, we can have any combinations of the above indicators. Considering the scope of processes from the operating system, such measures will influence in deciding about when to trigger the load balancing, which processes will be involved, and where are the destination places in which these processes will execute. Especially on the last topic, other factors to consider when selecting where to put a process include the nearness to resources, some processor and operating system capabilities, and specialized hardware/software features. We must first determine when to balance the load to turn the mechanism useful. Doing so is composed of two phases: (1) detecting that a load unbalancing exists and (2) determining if the cost of load balancing exceeds its possible benefits.

The use of load balancing in data science demands can vary depending on the structure of the parallel applications: Master-Slave, Bag of Tasks, Divide-and-Conquer, Pipeline, or Bulk-Synchronous Parallel [15, 16]. In the first two, we usually have a centralized environment where it is easy to know data about the whole set of resources, to dispatching tasks to them following their load and theoretical capacity. A traditional example of a combination of these parallel applications is the MapReduce framework. In the divide-and-conquer applications, we have a recursive nature to execute the parallel application where new levels of child nodes are created with the upper one cannot execute the tasks in an acceptable time interval. The challenge consists of dividing the tasks rightly following the capacity of the resources. Pipeline-based applications, in their turn, have a set of stages where each incoming task must cross. In order to maintain the cadence between the stages, they must execute in the same time interval, so an outcoming task from the stage n serves as the direct input for the stage $n + 1$. However, the fact of guaranteeing this capacity is not a trivial procedure because of the stages commonly present different complexities in terms of execution algorithms. Finally, bulk-synchronous applications are composed by supersteps, each one with local processing, and arbitrary communication and barrier phases. Load balancing is vital to guarantee that the slowest process does not compromise the performance of the entire application.

4. Open opportunities and trends

This section aims at compiling the previous two sections, so detailing open opportunities and trends when joining resource scheduling and load balancing and the area of data science. In this way, we compile these aspects as follows:

- Automatic transformation of a sequential data science demand to a parallel one—today data science executes locally to query databases and to build knowledge graphs. Sometimes these tasks are time consuming, then it is pertinent to transform a sequential demand in a parallel one to execute faster on multicore, multinode, and GPU architectures.

- Use of GPU cards as an accelerator for data science algorithms—write of data science demands that combine R and Python together with OpenCL or CUDA programming languages, so combining CPU and GPU codes with running fast and in parallel to address a particular data science demand.

- Combination of multimetric load balancing engine to handle data science efficiently—data science typically encompasses excellent access to IO (including main memory and hard disk) and a high volume of CPU cycles to process CPU-bound algorithms. In this way, the idea is to execute data science demands and learn their behavior, so proposing an adaptable load balancing metrics that take into account different parameters as input.

- Task migration heuristics—when developing long-running data science parallel codes, it is essential to develop task migration alternatives to reschedule demands from one resource to another. This is particularly pertinent on dynamic environments, either at the application or infrastructure level.

- Cloud elasticity to address data science demand—cloud elasticity comes to adapt the number of the resource following the current demand. Thus, we propose a combination of vertical and horizontal elasticity, together with reactive and proactive approaches to detect abnormal situations. We can use both consolidation and inclusion of resources, aiming to always accommodate the most appropriate number of resources for a particular and momentaneous data science demand.

- Definition of a standard API to deal with data science—frequently enterprises present several departments, each one with its data science demands. In this way, we envisage an opportunity on developing a standard framework (with a standard API too) to support the data science demands of the whole enterprise. The idea is to provide a dashboard with a collection of data science functions, also expressing the expected input and the output for each one.

- Smart correlation of events—enterprises regularly have timed data in several databases. We present an opportunity, at each time a problem is found, to take this particular timestamp and compare in the data sources looking for eventual data correlations. Thus, we can perceive relations such as: (1) if this happens, these other things will also happen and (2) this event happened because a set of prior events happened beforehand.

- Benchmark to evaluate a mapping of data science tasks to HPC resources—how we know if particular scheduling outperforms another one for executing a particular data science demand? We see as an opportunity for the exploration of benchmarks to evaluate scheduling and load balancing techniques that manipulate data science tasks. Thus, such benchmarks must define what they expect as input and provide a set of metrics as output. Yet, the output can be a single value, a collection of values (as a data vector), or a collection of elements of a data structure (e.g., timestamp and data are useful to develop user profiles and tracking of assets).

- Simulation environment to execute data science demands on distributed resources, but doing all of this a local program—simulation environments, like Simgrid or GridSim, are useful to use a sequential program to test and simulate complex parallel demands on a set of virtual resources. Thus, we can save time on testing different parameters and algorithms when developing scheduling and load balancing algorithms for data science.

- Definition of metrics to evaluate the scheduling and/or load balancing of data science tasks—CPU load, memory footprint, disk space, network throughput, and cache hit rate are examples of metrics that are commonly employed on distributed systems. Data science is a new area of knowledge, where we encourage the definition of new metrics to compare the execution of data science demands.

5. Conclusion

The continuous generation of data by different industry segments presents a valuable opportunity for analysis and knowledge extraction through data science methods. There is a high interest in studies that explore the application of data science to a variety of scenarios, each one with distinct characteristics that reflect on the composition of available datasets. Furthermore, there is not a single data science methodology that is applied to all possible data science problems. Consequently, the most common approach to data science problems is to define a sequence of methods that depend on the characteristics of the dataset and the intended results.

The constant growth in dataset sizes and the complexity of specific data science methods also impose a considerable challenge to provide the computational power required to process data and extract meaningful knowledge. In this context, cloud, fog, and grid computing architectures present themselves as ideal solutions to apply data science processes to massively sized datasets.

The distributed nature of such environments raises a series of new challenges, some of which widely studied in the literature. Nevertheless, the unique characteristics of data science workloads bring new aspects to these challenges, which require renewed attention from the scientific community.

This chapter focused on the specific challenge of scheduling and load balancing in the context of computational environments applied to data science. We presented an overview of data science processes, in addition to how scheduling and load balancing methodologies impact these processes and what aspects to consider when using distributed environments applied to data science. In particular, the challenge of enabling the automatic transformation of sequential data science demands into parallel ones is of particular interest because it abstracts part of the complexity involved in parallelizing data science tasks. As a result, such an automatic transformation promotes wider adoption of distributed environments as standard tools for large-scale data science processes.

Another notable challenge is to develop cloud elasticity techniques tailored to data science tasks. Such techniques must consider the specific requirements of data science processes to guarantee the proper reservation of resources and migration of tasks in order to guarantee a high throughput for such scenarios. These and the other investigated challenges represent prime research opportunities to increase the performance of data science processes.

Author details

Diórgenes Eugênio da Silveira[1], Eduardo Souza dos Reis[1], Rodrigo Simon Bavaresco[1], Marcio Miguel Gomes[1], Cristiano André da Costa[1], Jorge Luis Victoria Barbosa[1], Rodolfo Stoffel Antunes[1], Alvaro Machado Júnior[2], Rodrigo Saad[2] and Rodrigo da Rosa Righi[1*]

1 Universidade do Vale do Rio dos Sinos, São Leopoldo, RS, Brazil

2 DELL Computadores do Brasil, Eldorado do Sul, RS, Brazil

*Address all correspondence to: rrrighi@unisinos.br

References

[1] Amirian P, van Loggerenberg F, Lang T. Data Science and Analytics. Cham: Springer International Publishing; 2017. pp. 15-37

[2] Gibert K, Horsburgh JS, Athanasiadis IN, Holmes G. Environmental data science. Environmental Modelling and Software. 2018;**106**:4-12. Special Issue on Environmental Data Science. Applications to Air quality and Water cycle

[3] Grus J. Data Science from Scratch: First Principles with Python. 1st ed. O'Reilly Media, Inc.; 2015. Available from: https://www.amazon.com/Data-Science-Scratch-Principles-Python/dp/149190142X

[4] Ahmad A, Paul A, Din S, Rathore MM, Choi GS, Jeon G. Multilevel data processing using parallel algorithms for analyzing big data in high-performance computing. International Journal of Parallel Programming. 2018;**46**(3):508-527

[5] Bomatpalli T, Wagh R, Balaji S. High performance computing and big data analytics paradigms and challenges. International Journal of Computer Applications. 2015;**116**(04):28-33

[6] Singh D, Reddy CK. A survey on platforms for big data analytics. Journal of Big Data. 2014;**2**(1):8

[7] Dorr BJ, Greenberg CS, Fontana P, Przybocki M, Le Bras M, Ploehn C, et al. A new data science research program: evaluation, metrology, standards, and community outreach. International Journal of Data Science and Analytics. 2016;**1**(3):177-197

[8] Chasapis D, Moreto M, Schulz M, Rountree B, Valero M, Casas M. Power efficient job scheduling by predicting the impact of processor manufacturing variability. In: Proceedings of the ACM International Conference on Supercomputing (ICS'19). New York, NY, USA: ACM; 2019. pp. 296-307

[9] Liu L, Tan H, Jiang SH-C, Han Z, Li X-Y, Huang H. Dependent task placement and scheduling with function configuration in edge computing. In: Proceedings of the International Symposium on Quality of Service (IWQoS'19). New York, NY, USA: ACM; 2019. pp. 20:1-20:10

[10] Palyvos-Giannas D, Gulisano V, Papatriantafilou M. Haren: A framework for ad-hoc thread scheduling policies for data streaming applications. In: Proceedings of the 13th ACM International Conference on Distributed and Event-based Systems (DEBS '19). New York, NY, USA: ACM; 2019. pp. 19-30

[11] Feng Y, Zhu Y. PES: Proactive event scheduling for responsive and energy-efficient mobile web computing. In: Proceedings of the 46th International Symposium on Computer Architecture (ISCA '19). New York, NY, USA: ACM; 2019. pp. 66-78

[12] Topcuoglu H, Hariri S, Min-You WU. Performance-effective and low-complexity task scheduling for heterogeneous computing. IEEE Transactions on Parallel and Distributed Systems. 2002;**13**(3):260-274

[13] Menon H, Kale L. A distributed dynamic load balancer for iterative applications. In: Proceedings of the International Conference on High Performance Computing, Networking, Storage and Analysis (SC'13). New York, NY, USA: ACM; 2013. pp. 15:1-15:11

[14] Schepis L, Cuomo F, Petroni A, Biagi M, Listanti M, Scarano G. Adaptive data update for cloud-based internet of things applications. In:

Proceedings of the ACM MobiHoc Workshop on Pervasive Systems in the IoT Era (PERSIST-IoT'19). New York, NY, USA: ACM; 2019. pp. 13-18

[15] Bak S, Menon H, White S, Diener M, Kale L. Integrating openmp into the charm++ programming model. In: Proceedings of the Third International Workshop on Extreme Scale Programming Models and Middleware (ESPM2'17). New York, NY, USA: ACM; 2017. pp. 4:1-4:7

[16] Gandhi R, Liu HH, Hu YC, Lu G, Padhye J, Yuan L, et al. Duet: Cloud scale load balancing with hardware and software. SIGCOMM Computer Communication Review. 2014;**44**(4): 27-38

Chapter 7

Types of Task Scheduling Algorithms in Cloud Computing Environment

Tahani Aladwani

Abstract

Cloud computing is one of the most important technologies used in recent times, it allows users (individuals and organizations) to access computing resources (software, hardware, and platform) as services remotely through the Internet. Cloud computing is distinguished from traditional computing paradigms by its scalability, adjustable costs, accessibility, reliability, and on-demand pay-as-you-go services. As cloud computing is serving millions of users simultaneously, it must have the ability to meet all users requests with high performance and guarantee of quality of service (QoS). Therefore, we need to implement an appropriate task scheduling algorithm to fairly and efficiently meet these requests. Task scheduling problem is the one of the most critical issues in cloud computing environment because cloud performance depends mainly on it. There are various types of scheduling algorithms; some of them are static scheduling algorithms that are considered suitable for small or medium scale cloud computing; and dynamic scheduling algorithms that are considered suitable for large scale cloud computing environments. In this research, we attempt to show the most popular three static task scheduling algorithms performance there are: first come first service (FCFS), short job first scheduling (SJF), MAX-MIN. The CloudSim simulator has been used to measure their impact on algorithm complexity, resource availability, total execution time (TET), total waiting time (TWT), and total finish time (TFT).

Keywords: task scheduling algorithms, load balance, performance

1. Introduction

Cloud computing is a new technology derived from grid computing and distributed computing and refers to using computing resources (hardware, software, and platforms) as a service and provided to beneficiaries on demand through the Internet [1]. It is the first technology that uses the concept of commercial implementation of computer science with public users [2]. It relies on sharing resources among users through the use of the virtualization technique. High performance can be provided by a cloud computing, based on distributing workloads across all resources fairly and effectively to get less waiting time, execution time, maximum throughput, and exploitation of resources effectively. Still, there are many challenges prevalent in cloud computing, Task scheduling and load balance are the biggest yet because it is considered the main factors that control other performance criteria such as availability, scalability, and power consumption.

2. Tasks scheduling algorithms overview

Tasks scheduling algorithms are defined as the mechanism used to select the resources to execute tasks to get less waiting and execution time.

2.1 Scheduling levels

In the cloud computing environment there are two levels of scheduling algorithms:

- First level: in host level where a set of policies to distribute VMs in host.
- Second level: in VM level where a set of policies to distribute tasks to VM.

In this research we focus on VM level to scheduling tasks, we selected task scheduling algorithms as a research field because it is the biggest challenge in cloud computing and the main factor that controls the performance criteria such as (execution time, response time, waiting time, network, bandwidth, services cost) for all tasks and controlling other factors that can affect performance such as power consumption, availability, scalability, storage capacity, buffer capacity, disk capacity, and number of users.

2.2 Tasks scheduling algorithms definition and advantages

Tasks scheduling algorithms are defined as a set of rules and policies used to assign tasks to the suitable resources (CPU, memory, and bandwidth) to get the highest level possible of performance and resources utilization.

2.2.1 Task scheduling algorithms advantages

- Manage cloud computing performance and QoS.
- Manage the memory and CPU.
- The good scheduling algorithms maximizing resources utilization while minimizing the total task execution time.
- Improving fairness for all tasks.
- Increasing the number of successfully completed tasks.
- Scheduling tasks on a real-time system.
- Achieving a high system throughput.
- Improving load balance.

2.3 Tasks scheduling algorithms classifications

Tasks scheduling algorithms classified as in **Figure 1**.

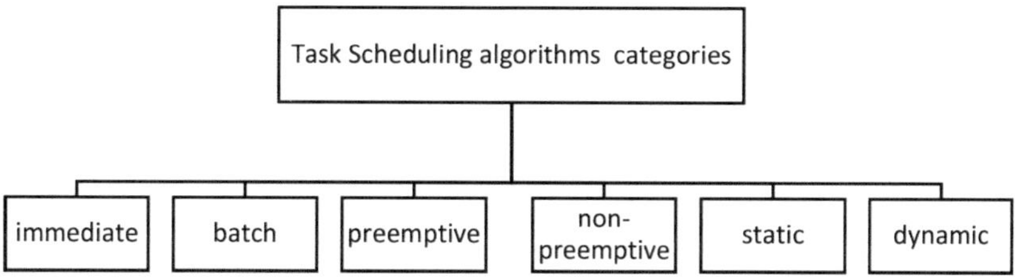

Figure 1.
Tasks scheduling classes.

2.3.1 Tasks scheduling algorithms can be classified as follows

- Immediate scheduling: when new tasks arrive, they are scheduled to VMs directly.
- Batch scheduling: tasks are grouped into a batch before being sent; this type is also called mapping events.
- Static scheduling: is considered very simple compared to dynamic scheduling; it is based on prior information of the global state of the system. It does not take into account the current state of VMs and then divides all traffic equivalently among all VMs in a similar manner such as round robin (RR) and random scheduling algorithms.
- Dynamic scheduling: takes into account the current state of VMs and does not require prior information of the global state of the system and distribute the tasks according to the capacity of all available VMs [4–6].
- Preemptive scheduling: each task is interrupted during execution and can be moved to another resource to complete execution [6].
- Non-preemptive scheduling: VMs are not re-allocated to new tasks until finishing execution of the scheduled task [6].

In this research, we focus on the static scheduling algorithms. Static scheduling algorithm such as first come first service (FCFS), shortest job first (SJF), and MAX-MAX scheduling algorithms in complexity and cost within a small or medium scale.

2.4 Task scheduling system in cloud computing

The task scheduling system in cloud computing passes through three levels [7].

- The first task level: is a set of tasks (Cloudlets) that is sent by cloud users, which are required for execution.
- The second scheduling level: is responsible for mapping tasks to suitable resources to get highest resource utilization with minimum makespan. The makespan is the overall completion time for all tasks from the beginning to the end [7].
- The third VMs level: is a set of (VMs) which are used to execute the tasks as in **Figure 2**.

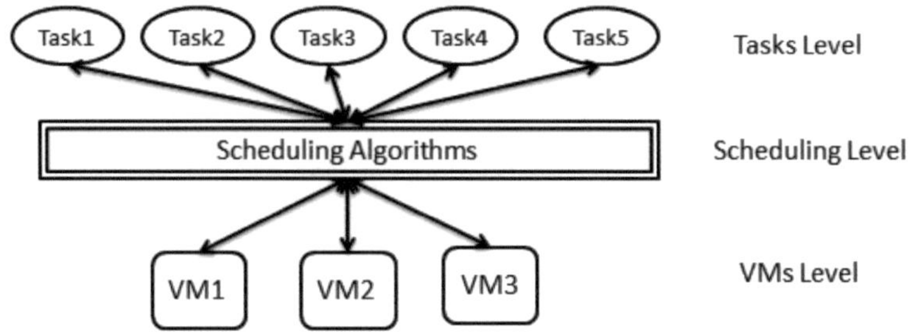

Figure 2.
Task scheduling system.

2.5 This level passes through two steps

- The first step is discovering and filtering all the VMs that are presented in the system and collecting status information related to them by using a datacenter broker [8].
- In the second step a suitable VM is selected based on task properties [8].

3. Static tasks scheduling algorithms in cloud computing environment

3.1 FCFS

FCFS: the order of tasks in task list is based on their arriving time then assigned to VMs [3].

3.1.1 Advantages

- Most popular and simplest scheduling algorithm.
- Fairer than other simple scheduling algorithms.
- Depend on FIFO rule in scheduling task.
- Less complexity than other scheduling algorithms.

3.1.2 Disadvantages

- Tasks have high waiting time.
- Not give any priority to tasks. That means when we have large tasks in the begin tasks list, all tasks must wait a long time until the large tasks to finish.
- Resources are not consumed in an optimal manner.
- In order to measure the performance achieved by this method, we will be testing them and then measuring its impact on (fairness, ET, TWT, and TFT).

3.1.3 Assumptions

Some of the assumptions must be taken into account when scheduling tasks to VMs in the cloud computing environment.

- Number of tasks should be more than the number of VMs, which means that each VM must execute more than one task.
- Each task is assigned to only one VM resource.
- Lengths of tasks varying from small, medium, and large.
- Tasks are not interrupted once their executions start.
- VMs are independent in terms of resources and control.
- The available VMs are of exclusive usage and cannot be shared among different tasks. It means that the VMs cannot consider other tasks until the completion of the current tasks is in progress [3].

Tasks lengths: assume we have 15 tasks with their lengths as in **Table 1**.

Task	Length
t1	100000
t2	70000
t3	5000
t4	1000
t5	3000
t6	10000
t7	90000
t8	100000
t9	15000
t10	1000
t11	2000
t12	4000
t13	20000
t14	25000
t15	80000

Table 1.
Set of tasks with different length orders depends on the arrival time for each task.

3.1.4 VM properties

Assume we have six VMs with different properties based on tasks size:

$$\text{VM list} = \{VM1, VM2, VM3, VM4, VM5, VM6\}.$$

$$\text{MIPS of VM list} = \{500, 500, 1500, 1500, 2500, 2500\}.$$

We selected a set of VMs with different properties to make each category have VMs with appropriate ability to serve a specific class of tasks, to improve the load balance. Because when we use VMs with same properties with all categories it leads to load imbalance, where each class is different from other classes in terms of tasks lengths.

3.1.5 When applying FCFS, work mechanism will be as following

Figure 3 shows FCFS tasks scheduling algorithm working mechanism and how tasks are executed based on their arrival time.

Dot arrows refer to first set of tasks scheduling based on their arrival time.

Dash arrows refer to second set of tasks scheduling based on their arrival time.

Solid arrows refer to third set of tasks scheduling based on their arrival time.

And here it is clear to us that t1 is too large compared with t7 and t12. However, t7 and t12 must wait for t1, which leads to an increase in the TWT, ET, TFT, and a decrease in fairness.

$$VM1 = \{t1 \longrightarrow t7 \longrightarrow t12\}.$$

$$VM2 = \{t2 \longrightarrow t8 \longrightarrow t14\}.$$

$$VM3 = \{t3 \longrightarrow t9 \longrightarrow t15\}.$$

$$VM4 = \{t4 \longrightarrow t10\}.$$

$$VM5 = \{t5 \longrightarrow t11\}.$$

$$VM6 = \{t6 \longrightarrow t13\}.$$

Table 2 shows how FCFS scheduling algorithm increases waiting time for all tasks.

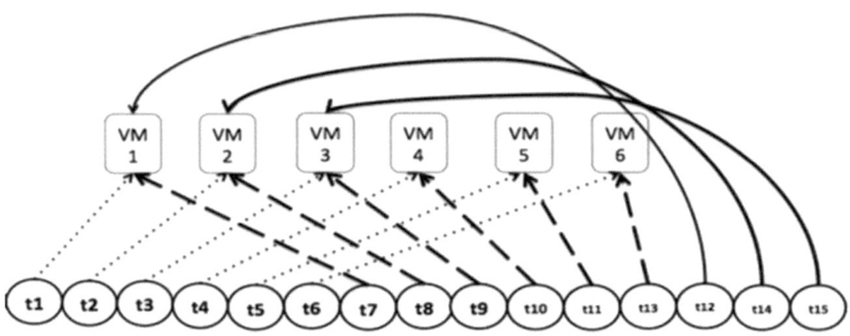

Figure 3.
FCFS work mechanism.

DOI: http://dx.doi.org/10.5772/intechopen.86873

Task	ET	Waiting time		
t1	200	VM1		
t2	140	VM2		
t3	3.33	VM3		
t4	0.66	VM4		
t5	1.2	VM5		
t6	4	VM6		
t7	180	Wait(200)	VM1	
t8	200	Wait(140)	VM2	
t9	10	Wait(3.33)	VM3	
t10	0.66	Wait(0.66)	VM4	
t11	0.8	Wait(1.2)	VM5	
t12	1.6	Wait(4)	VM6	
t13	40	Wait(380)		VM1
t14	50	Wait(340)		VM2
t15	53.33	Wait(13.33)		VM3

Table 2.
Waiting times of tasks in FCFS.

3.2 SJF

Tasks are sorted based on their priority. Priority is given to tasks based on tasks lengths and begins from (smallest task ≡ highest priority).

3.2.1 Advantages

- Wait time is lower than FCFS.
- SJF has minimum average waiting time among all tasks scheduling algorithms.

3.2.2 Disadvantages

- Unfairness to some tasks when tasks are assigned to VM, due to the long tasks tending to be left waiting in the task list while small tasks are assigned to VM.
- Taking long execution time and TFT.

3.2.3 SJF work mechanism

When applying SJF, work mechanism will be as follows:

Assume we have 15 tasks as in **Table 1** above. We will be sorting tasks in the task list, as in **Table 3**. Tasks are sorted from smallest task to largest task based on their lengths as in **Table 3**, then assigned to VMs list sequential.

3.2.4 Execute tasks will be

$$VM1 = \{t4 \longrightarrow t6 \longrightarrow t7\}.$$

$$VM2 = \{t10 \longrightarrow t9 \longrightarrow t1\}.$$

$$VM3 = \{t11 \longrightarrow t13 \longrightarrow t8\}.$$

$$VM4 = \{t5 \longrightarrow t14\}.$$

$$VM5 = \{t12 \longrightarrow t2\}.$$

$$VM6 = \{t3 \longrightarrow t15\}.$$

Table 4 shows that the large tasks must be waiting in the task list until the smallest tasks finish execution.

3.3 MAX-MIN

In MAX-MIN tasks are sorted based on the completion time of tasks; long tasks that take more completion time have the highest priority. Then assigned to the VM with minimum overall execution time in VMs list.

3.3.1. Advantages

- Working to exploit the available resources in an efficient manner.
- This algorithm has better performance than the FCFS, SJF, and MIN-MIN algorithm.

3.3.2 Disadvantages

- Increase waiting time to small and medium tasks; if we have six long tasks, in MAX-MIN scheduling algorithm they will take priority in six VMs in VM list, and short tasks must be waiting until the large tasks finish.

Unfairness to some or most small and medium tasks when tasks are assigned to VM.

- When applying MAX-MIN, Work Mechanism will be as follows.

Tasks	*t4*	*t10*	*t11*	*t5*	*t12*	*t3*	*t6*	*t9*	*t13*	*t14*	*t2*	*t15*	*t7*	*t1*	*t8*
lengths	1000	1000	2000	3000	4000	5000	10000	15000	20000	25000	70000	80000	90000	100000	100000

Table 3.
A set of tasks sorted based on SJF scheduling algorithm.

Task	ET	Waiting time		
t4	2	VM1		
t10	2	VM2		
t11	1.33	VM3		
t5	2	VM4		
t12	1.6	VM5		
t3	2	VM6		
t6	20	Wait(2)	VM1	
t9	30	Wait(2)	VM2	
t13	13.33	Wait(1.33)	VM3	
t14	16.66	Wait(2)	VM4	
t2	28	Wait(1.6)	VM5	
t15	32	Wait(2)	VM6	
t7	180	Wait(22)		VM1
t1	200	Wait(32)		VM2
t8	66.66	Wait(14.66)		VM3

Table 4.
Waiting times of tasks in SJF.

Tasks	t1	t8	t7	t15	t2	t14	t13	t9	t6	t3	t12	t5	t11	t10	t4
lengths	100000	100000	90000	80000	70000	25000	20000	15000	10000	5000	4000	3000	2000	1000	1000

Table 5.
A set of tasks sorted based on MAX-MIN scheduling algorithm.

Assume we have 15 tasks as in **Table 1** above. We will be sorting tasks in task list as in **Table 5**. Tasks sorted from largest task to smallest task based on highest completion time. They are then assigned to the VMs with minimum overall execution time in VMs list.

3.3.3 Execute tasks will be

$$VM6 = \{t1 \longrightarrow t13 \longrightarrow t11\}.$$

$$VM5 = \{t8 \longrightarrow t9 \longrightarrow t10\}.$$

$$VM4 = \{t7 \longrightarrow t6 \longrightarrow t4\}.$$

$$VM3 = \{t15 \longrightarrow t3\}.$$

$$VM2 = \{t2 \longrightarrow t12\}.$$

$$VM1 = \{t14 \longrightarrow t5\}.$$

Tables 6 and **7** shows that the small and medium tasks must be waiting in the task list until the large tasks finish execution.

Figure 4 shows the TWT and TFT for the three tasks scheduling algorithms FCFS, SJF, and MAX-MIN. SJF tasks scheduling algorithm is the best in term of TWT and TFT.

Task	ET	Waiting time		
t1	40	VM6		
t8	40	VM5		
t7	60	VM4		
t15	53.33	VM3		
t2	140	VM2		
t14	50	VM1		
t13	8	Wait(40)	VM6	
t9	6	Wait(40)	VM5	
t6	6.66	Wait(60)	VM4	
t3	3.33	Wait(53.33)	VM3	
t12	8	Wait(140)	VM2	
t5	6	Wait(50)	VM1	
t11	0.8	Wait(48)		VM6
t4	0.4	Wait(46)		VM5
t10	0.67	Wait(66.67)		VM4

Table 6.
Waiting time of tasks in MIX-MIN scheduling algorithm.

	FCFS	SJF	MAX-MIN
TWT	739.19	79.59	404
TFT	1969.69	678.69	968.698

Table 7.
Comparison between FCFS tasks scheduling algorithm, SJF, and MAX-MIN in terms of TWT and TFT.

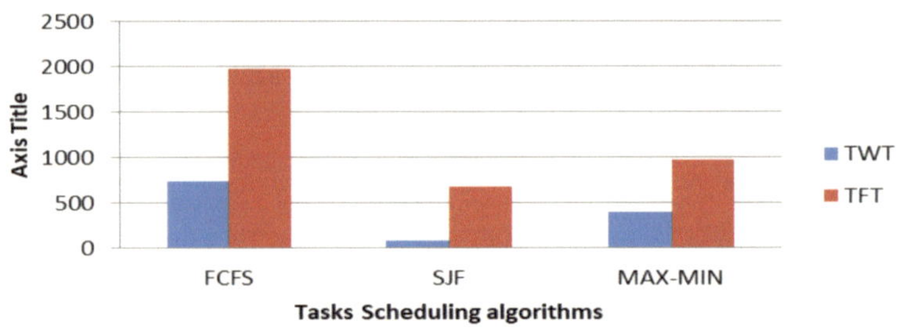

Figure 4.
Comparison between FCFS tasks scheduling algorithm, SJF, and MAX-MIN in terms of TWT and TFT.

4. Conclusion

This chapter introduces the meaning of the tasks scheduling algorithms and types of static and dynamic scheduling algorithms in cloud computing environment. This chapter also introduces a comparative study between the static task scheduling algorithms in a cloud computing environment such as FCFS, SJF, and MAX-MIN, in terms of TWT, TFT, fairness between tasks, and when becoming suitable to use?

Experimentation was executed on CloudSim, which is used for modeling the different tasks scheduling algorithms.

Author details

Tahani Aladwani
Mecca, Saudi Arabia

*Address all correspondence to: aladwani_tahani@yahoo.com

References

[1] Ramotra A, Bala A. Task-Aware Priority Based Scheduling in Cloud Computing [master thesis]. Thapar University; 2013

[2] Microsoft Azure website. [Accessed: 01 October 2017]

[3] Kumar Garg S, Buyya R. Green Cloud Computing and Environmental Sustainability, Australia: Cloud Computing and Distributed Systems (CLOUDS) Laboratory Department of Computer Science and Software Engineering, The University of Melbourne; 2012

[4] Al-maamari A, Omara F. Task scheduling using PSO algorithm in cloud computing environments. International Journal of Grid Distribution Computing. 2015;**8**(5):245-256

[5] http://www.pbenson.net/2013/04/the-cloud-defined-part-1-of-8-on-demand-self-service/ [Accessed: 01 October 2017]

[6] Endo P, Rodrigues M, Gonçalves G, Kelner J, Sadok D, Curescu C. High availability in clouds: Systematic review and research challenges. Journal of Cloud Computing Advances, Systems and Applications. 2016

[7] http://www.techinmind.com/what-is-cloud-computing-what-are-its-advantages-and-disadvantages/ [Accessed: 01 October 2017]

[8] https://siliconangle.com/blog/2016/04/29/survey-sees-rapid-growth-in-enterprise-cloud-adoption/ [Accessed: 01 October 2017]